Journey to the Center of Our Galaxy

A Voyage in Space and Time

JOEL DAVIS

CB

CONTEMPORARY
BOOKS

CHICAGO

Library of Congress Cataloging-in-Publication Data

Davis, Joel, 1948–
    Journey to the center of our galaxy : a voyage in space and time /
Joel Davis.
      p.   cm.
    Includes bibliographical references (p. 321) and index.
    ISBN 0-8092-4242-7 (cloth)
        0-8092-3847-0 (paper)
    1. Milky Way.   I. Title.
QB857.7.D38   1991
523.1'13—dc20   91-22713
                                                  CIP

Illustrations from *Contemporary Astronomy*, third edition, by Jay
Pasachoff, copyright © 1985 by Saunders College Publishing, a division
of Holt, Rinehart and Winston, Inc., reprinted by permission of the
publisher.

Front cover photographs, clockwise from upper left: diagram of six
astronomical systems from Kircher's *Inter extaticum coeleste*; Johannes
Kepler; Andromeda Galaxy, © National Optical Astronomy
Observatories; the celestial spheres, from Aspian's *Cosmographia*;
William Herschel; radio image of the Galactic Center Jet, courtesy of
Yoshiaki Sofue; Sir Isaac Newton.

Cover design by Georgene Sainati

Published by Contemporary Books, Inc.
180 North Michigan Avenue, Chicago, Illinois 60601
Manufactured in the United States of America
International Standard Book Number: 0-8092-4242-7 (cloth)
                                            0-8092-3847-0 (paper)

This one is for my sister and brothers:
Chris, Mickey, Marie, Tim, Peter, and Ed

CONTENTS

ACKNOWLEDGMENTS

This book would not have been possible without the help of many people and organizations. My thanks to Isaac Asimov, Dr. Gregory Benford, Dr. Bruce Carney, Dr. Thomas M. Dame, Dr. Laura Danly, Dr. K. C. Freeman, Dr. Ian Gatley, Martha Hazen, Dr. Paul Hodge, Dr. Bruce Margon, Dr. Mark Morris, Dr. Bruce Moskowitz, Larry Niven, Dr. Yoshiaki Sofue, Dr. Patrick Thaddeus, Dr. Keven Uchida, Dr. J. P. Vallee, and Dr. Rosemary F. G. Wyse, and to the Astronomical Society of the Pacific, Cerro Tololo Inter-American Observatory, the Hansen Planetarium, the Harvard College Archives and Plate Stacks, Kitt Peak National Observatory, Lick Observatory and the Regents of the University of California, the National Radio Astronomy Observatory, and to Saunders College Publishing, Inc. I want especially to thank Isaac Asimov, Greg Benford, and Larry Niven for their gracious permission to quote from their science fiction stories (first a book introduction, Isaac, and now a *Foundation* fragment; what next?); and Greg Benford and Mark Morris for reading through an early draft of this book's manuscript and offering valuable suggestions and corrections. Any errors of fact herein are not their fault, but mine.

Thanks, also, to Greg and Astrid Bear; Joshua Bilmes; Dr. John Cramer; Gerald and Toni Davis; Elana Freeland; Therese Gonneville, OSB; all of the people of the Nisqually Dance Circle; Dean and Carla Jones and the folks at the Four Seasons Bookstore; Carol Knobel; the librarians at The Evergreen State College Library, the Timberland (Washington) Library System, the University of Washington Astronomy/Physics Library, and the Washington State Library; Megan Lindholm; Vonda N. McIntyre; Mary Giles Mailhot, OSB; Barbara Park; Gary Sandwick; the community of St. Placid Priory; Carol Stilz; and the students in my writing classes at The Evergreen State College and the University of Washington.

Finally, deepest thanks to the Tuesday noon CoDA group, the Friday ACA 12 Step Workbook Family, the Sunday afternoon ACA/Alanon Family Group, and to the One who helps me make it one day at a time.

To those I have inadvertently omitted, my apologies and my thanks.

INTRODUCTION

From our location on Earth, we cannot see the center of our Galaxy with our unaided eyes. But we can get a dazzling glimpse of at least parts of this star city in which we live.

Some cloudless evening, drive to a sparsely populated place outside your city or town, a place far from homes, office buildings, and streetlights. Take a lawn chair or a blanket with you. Sit or lie down with your face to the sky above and your head to the north. Look up and south.

When your eyes have adapted to the darkness, you will see a faint river of light stretching across the heavens. That's the Milky Way. It's nothing more or less than our own Galaxy seen from the inside, at a point about twenty-nine thousand light-years from its center. Near its southern end, in the constellation Sagittarius, the Milky Way thickens like spilt cream. Those dense star clouds you see lie close to the center of our Galaxy. That, however, is the best view you will get of the Galactic center from this planet at this point in our 250-million-year orbital ride around the Galaxy.

The actual Galactic core is hidden from our view by vast intervening clouds of dark gas and dust. They look like black rifts or holes in the Sagittarius star clouds. They are not holes, but shutters. Humans see only visible light, a small part of the total spectrum. Visible light from the center of our Galaxy cannot penetrate those dust clouds. Only in the last two decades have we been able to actually "see" the center of the Galaxy, using instruments that detect other wavelengths. Telescopic eyes that perceive gamma rays, x-rays, infrared (heat) radiation, and radio waves have finally made it possible for humans to actually see what lies at the Galactic core.

And what do we find there? We're still not entirely sure, but all the evidence seems to point to the presence of a Monster: a giant black hole that is literally gobbling up the Galaxy.

A hundred years ago we did not even know the true shape of this city of stars, this Milky Way Galaxy, in which we live. Neither did we know its size, nor our location within it. In fact, a century ago we did not even know the nature of light or gravity, how stars are born and die, or what powers them. Fifty years ago we had started getting the first good answers to those questions. However, we still had no idea what the *center* of the Galaxy was like. Today we are beginning to get a good look. *Journey to the Center of Our Galaxy* is a voyage in both time and space, in which we will discover the nature of stars, of the Galaxy, of light and gravity, and of the bizarre object that lies at the center of this celestial whirlpool of stars, dust, gas, and who knows what else. Like the intrepid adventurers in Jules Verne's famous novel *Journey to the Center of the Earth*, we will encounter marvelous things, sights to astonish and humble us.

Our journey is one of imagination, history, and scientific discovery. We will begin in imagination, looking over the shoulders of three eminent science fiction writers who have conjured up their own images of the center of the Galaxy. Like every good science fiction story, theirs are based on scientific facts—as understood at the time they wrote their stories. So our three quick surveys will also give us an overview of how the scientific conceptions of the Galactic center have changed in the last half-century.

Journey to the Center of Our Galaxy is also a journey in time. For the voyage to the center of the Galaxy is a voyage upon the ocean of the human mind and heart. It is an ocean whose surface has been tossed by the winds of thought and imagination, whose depths are roiled by currents of philosophy, religion, and mythology. We do not view the world today as our ancestors did. Even those who lived a mere three hundred to four hundred years ago perceived reality in a fashion far different from us. And those humans who lived four thousand and more years in the past had worldviews radically different from our own. Our perceptions and assumptions about "what is at the center of the universe" have changed, and the story of those changes is an essential part of this book.

This is also the story of our gradual understanding of an elemental force of nature. We call it gravity. The current hypothesis of the presence of an all-devouring black hole at the center of the Galaxy is due to our understanding of what gravity is, of how it works, and of how nothing can escape its grasp, if it is strong enough.

We will therefore travel from the distant past to our present day, from ancient myths about the creation of the universe and what lies at its center—namely, us, sometimes on the back of a turtle—to today's picture of the Galaxy as one of trillions, in a universe of unimaginable size that itself has no center at all. Our journey through time to the center of the Galaxy will take us from ancient Sumer to Greece, from the deserts of the American Southwest to the jungles of equatorial South America, from first-century B.C. Egypt to sixteenth-century Poland, from the Italy of Galileo and Cardinal Bellarmine to nineteenth-century England and the German-born astronomer William Herschel.

From there we will travel to the turn of the twentieth century and the study of a despairing German physicist named Planck, then to the Swiss Patent Office and the desk of another, not-so-despairing German named Einstein. We will journey to a lecture hall in Boston in November 1920 and to wartime blacked-out Los Angeles in 1944. This part of our journey will eventually take us to astronomical observatories around the world during the 1950s and 1960s, then to the edge of space and the first Earth-orbiting space telescopes of the 1970s and 1980s. Finally, we will come to the present and gaze into the heart of the Galaxy using an instrument that lies spread across miles of New Mexico desert—the ancient home of a people who even today view the cosmos in a manner far different from that of us "sophisticated" scientific Westerners.

Finally, we will embark upon a "physical" voyage into the

very center of the Milky Way, to the edge of a great black hole that may be "eating" the Galaxy. We will travel from the outer fringes of the Galaxy, three hundred thousand light-years from the center, into the very core of the Milky Way. One light-year is 5.88 trillion miles, the equivalent of 236 million trips around the Earth. Three hundred thousand light-years is 1.7 quintillion miles. That's 1,700,000,000,000,000,000 miles.

According to Einstein's special theory of relativity, nothing in "the real world" can accelerate to a velocity faster than that of light. Even light itself would take three hundred thousand years to make the journey we are about to make. However, few people realize that special relativity *does not explicitly forbid travel faster than light*. The only trick is getting past the light-speed barrier in the first place. Once past it, though, we can go as fast as we wish. For example, special relativity itself does not forbid the existence of "tachyons," hypothetical subatomic particles that never travel slower than the speed of light.

More to our particular needs here, special relativity also does not apply to flights of imagination. So, for the sake of brevity and in the name of literary license, we will make this journey to the center of the Milky Way Galaxy in a fictional spaceship that travels faster than light by means of "Jumps." With the judicious application of some science fictional sleight of hand (which some may prefer to call scientific double-talk, and others imaginative playfulness), our "metaship" will skip across space and time the way a specially thrown stone skips across the surface of a pond.

These Jumps will take us from mystery to mastery. On the way we will travel through the mysterious "dark matter" corona of our Galaxy; past globular clusters of millions of stars that orbit the Galaxy like vast stellar cities in space; through clouds of molecular gas and dust that are the breeding grounds of new stars; through the expanding shells of supernovae, stars that have exploded and turned into neutron stars; and deep into the central region of the Milky Way. There we will encounter two enormous "smoke rings" of gas that were blown out of the Galactic core by explosions that took place only a few tens of millions of years ago. Still closer to the center, we will encounter gargantuan rivers of ionized gas, called plasma, streaming toward the center. Finally, we will gaze into the face of the Monster itself and its immediate surroundings.

William Shakespeare once wrote that even "time . . . must have a stop." It is true that the nature of black holes prevents us from ever knowing what lies within them, what is going on there. No information can ever escape from their overwhelming gravitational grasp. At the edge of that massive black hole at the center of our Galaxy, our journey in time and space will end.

But not, I hope, our journey of imagination. As long as we remain human and grow, and wonder, and ask questions, that journey will continue.

PART I
A JOURNEY THROUGH
IMAGINATION

But thought's the slave of time, and life time's fool;
And time, that takes survey of all the world,
Must have a stop. . . .
  —Hotspur, in *Henry IV, Part I*, by William Shakespeare

—Oh my God!—It's full of stars!
  —Dave Bowman, in *2001: A Space Odyssey*,
  by Arthur C. Clarke

1
FROM TRANTOR TO TIDES OF LIGHT

We are standing in a place of unimaginable energy and vio-lence. Were it not for the force fields surrounding us, fields that would seem like magic to our twentieth-century ancestors, we would have been dead long before we arrived here.

"Here" is a desolate chunk of rock, blasted for millennia by gouts of matter and energy spiraling in from beyond and spouted out from within.

"Here" is near the very center of the Milky Way Galaxy—the home galaxy, home of Sol, of Earth, of the human species. We are only 2.8 trillion kilometers out, just three-tenths of a light-year from the center.

We look up. Our senses are highly augmented. We can see deep into the infrared, microwave, and radio bands of the spec-trum and high into the ultraviolet and x-ray regions beyond blue and violet. Our ears can "hear" energies that our great-grand-parents could only trace on computer screens. We can taste molecular hydrogen (a little like lemon-drop soup) and mole-cules of carbon monoxide (sharp and biting, like hot peppers). We tune out everything but sight. Most of the sky is blazing with wreaths of color, like gargantuan veils of blue, green, red, orange-pink.

In the midst of the swirl of colors and sounds and tastes is a small dark disk, blacker than black, the blackness of emptiness, blacker even than the blackest and emptiest intergalactic space we saw at the beginning of our journey. Cutting across that dark disk, like the line through the ancient Greek letter theta (Θ) is a line of deep red fire. We are looking edge-on at an accretion disk, *fed by dust and gas that spiral in from all the muck surrounding our dead asteroid and from beyond. And the accretion disk itself encircles the dark place in front of us. It looks tiny from where we stand, but it is not.*

It is the Monster at the center of the Galaxy. It is a black hole with the mass of a million suns. It is slowly consuming the Galaxy. And it is what we have come three hundred thousand light-years to see.

We live today at the end of a millennium, in the middle of a sea change in the way our species perceives the cosmos and our relationship to it, and at the beginning of an era of unprecedented exploration. Our time, like all time, is like a bubble expanding outward into infinity. We have always lived on the edge of that bubble. We have always wondered what lies beyond it and behind us.

Our time is also a time of chaos, of fragmentation, of scattered consciousness. Unlike our ancestors who lived before the invention of writing, cities, and agriculture, we have no *center* to our lives. The myths and mythic symbols that were the hub of past cultures and societies were overthrown and replaced by the myth and symbols of science. The central symbol of that science was the Newtonian worldview. The universe, this worldview held, is a Great Watch, constructed by a Great Watchmaker, who turned it on and let it begin ticking away. It is perfect, immutable, endless.

Finally, in the memory of some still living, that worldview itself was destroyed by the work of three men: the trinity of Planck, Einstein, and Bohr. Max Planck showed that light is neither a particle nor a wave, but something that acts like both. Einstein's special theory of relativity demolished the absoluteness of space and time and showed that—contrary to poet T. S. Eliot—there is no "still point of the turning world." His general theory of relativity demonstrated that not even gravity and light can be absolutely trusted. Gravity is nothing more than a warping of space and time, beams of light can be twisted, and there is no such thing as a straight line. Niels Bohr, stepping forward from the work of Planck, constructed the theory of quantum mechanics. This theory of physics (the most predictively accurate ever created) essentially says that, for all practical purposes nothing exists until it is observed, or until it leads to some observable phenomenon, and that some characteristics of the cosmos can never—*never*—be known.

One of the interpretations of quantum mechanics (and there are many, each of which is no more or less valid than the others) suggests that not even the universe is unique. This "many-worlds" interpretation asserts that with each observation, each choice made by a self-aware sentient being, an entirely new universe comes into existence. Another implication of the "many-worlds" interpretation is that an infinity of universes has *always* existed.

It seems that at least one common thread runs through these changes in worldviews over the last thousand years or so. It has to do with *the center*: both the center of our spiritual and psychological lives and the center of the universe. Our perception of what is the center of the cosmos has changed. Many prescientific myths present the Earth as the center of the universe. With the scientific revolution in the sixteenth and seventeenth centuries came a new perception of the sun as the center of the universe. And now, as the Millennium approaches, we know that the sun and our solar system do not lie at the center. In fact, it is the prevailing scientific view (call it myth if you wish; there is no shame in doing so, for as Chapter 2 explains, myth does *not* mean "falsehood") that *there is no center to the universe*, to the cosmos, to anything at all. Our universe expands, but from no fixed central point. Indeed, our universe may be one of an infinity of universes.

Philosophers used to speak of "the plurality of worlds." Today we speak of the plurality of universes.

Yet we remain fascinated with the idea of a center. A center is important to us. It is vital to us, in the literal sense of "vital." Without a center, there is no life—certainly no emotional, psychological, or spiritual life, even if there is physical life. Many psychologists and counselors remark on the absolute necessity of being in touch with "that which lies at our center"—our

Inner Self, Inner Child, Child Within, Divine Spark, Christ Consciousness, Buddha Nature, Tao; it has many names. When we are not connected to our Center, we are lost people, stumbling

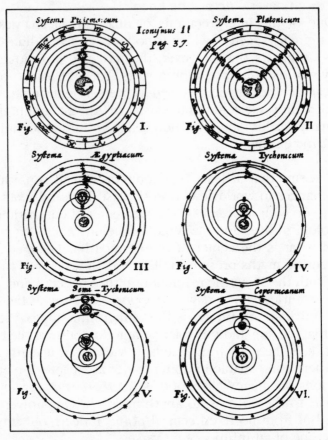

This diagram of six astronomical systems or worldviews was published in 1660. The first five are geocentric, with the Earth at the center of the cosmos. The only difference between the Ptolemaic and Platonic systems (I and II) is that the Ptolemaic system puts the sun's orbit beyond that of Mercury and Venus, while the Platonic system places it within Mercury's orbit. In the Egyptian system (III), Mercury and Venus orbit the sun, which in turn orbits the Earth. The Tychonic system (IV) places Jupiter and Saturn in orbit around the sun (which is still orbiting the Earth), while the semi-Tychonic system does not. The sixth astronomical system in the illustration is the Copernican system, which places the sun at the center of the cosmos and the Earth orbiting it.

Illustration from Michael J. Crowe, *Theories of the World from Antiquity to the Copernican Revolution*, Dover Publications, Inc., 1990. Reprinted by permission of the publisher.

through our daily lives and fighting off despair as best we can.

That desire to know our own metaphysical or spiritual center, and to be connected to it, is mirrored by the desire to know more about other, more physical centers. The cosmos may not have a center, but every object within it does. Asteroid, satellite, Earth, sun, solar system, globular cluster, galaxy, galactic supercluster, each has a center. We don't have to be astronomers or astrophysicists to wonder about and be fascinated by what might lie at the center of the Earth, or what the center of the sun is like, or what might dwell at the center of our Milky Way Galaxy.

The French writer Jules Verne spun a marvelous yarn about our Earth's center in his classic novel *Journey to the Center of the Earth*. Today geophysicists construct maps of the "hills" and "valleys" in the ultramolten iron-nickel core of our planet. Not long ago physicist and science fiction writer Dr. Robert Forward wrote a novel called *Dragon's Egg*, which suggests that the center of the sun contains tiny black holes. Indeed, physicists do speculate about what is going on in the sun's center. They are puzzled by a mysterious lack of subatomic particles called neutrinos. They should be pouring out of the sun, but they come out only as a trickle.

Since 1987 astronomers have been gazing through their telescopes at an object so far away that its light took more than 160,000 years to arrive. A supergiant star collapsed into itself, with some of its matter rebounding from the center into the explosion called a supernova. Supernova 1987A, as it is now called, has given scientists new information about what happens at the center of a star as it dies in this spectacular fashion.

And what about the center of our Galaxy? Our journey into that region has been a long one, in both space and time. The first crude outlines of the Galaxy were puzzled out only about 150 years ago. Only in the last fifty years or so have we even discovered the detailed structure of the Milky Way, with its arms, corona, halo, giant molecular clouds, vast gas and dust rings, and attendant globular clusters. There is still the enormous mystery of the Galaxy's "missing mass," the more than 90 percent of the Galaxy that is apparently some form of "dark matter."

Finally, in the last decade, we have begun to "see" the central region of the Milky Way. And what we find there, at the center of our Galaxy, is a mystery that may be a Monster.

We have been there before, in the imaginations of writers. The ancient Greeks had their own visions of what lies at the center of the cosmos, and we will soon sample the cosmological speculations of Anaxagoras, Anaximander, Aristotle, Eudoxus, Hipparchus, Homer, Plato, and Pythagoras.

But even in this age of science and rationality, of galaxies and black holes and infinite universes, some writers have sailed on the winds of imagination into a region that no one has yet seen with the naked eye. The descriptions with which they have returned have differed from one another. And that is not too surprising. The literary genre we call science fiction has its roots in the scientific knowledge of a particular writer's particular place and time. An SF writer of the 1940s would naturally conjure up a different vision of the Galactic core than one writing in the late 1980s. Here, then, are three different looks at the same place, the center of the Milky Way Galaxy, seen from three different vantage points: the 1940s, the 1960s, and the 1980s.

The Center of the Galaxy I: Trantor

On August 11, 1941, a twenty-one-year-young science fiction writer named Isaac Asimov began writing a short story called "Foundation." It was the story of the collapse of an immense Galactic empire, and of one man's plan to shorten the inevitable Dark Ages to follow. In this story Asimov invented a new word (psychohistory) and an enduring fictional character named Hari Seldon. Seldon is a psychohistorian who uses his scientific knowledge of psychology and history to anticipate the various crises that will confront the people struggling to save humanity from chaos. Seldon sets up two "Foundations," located on planets "at opposite ends of the Galaxy." These Foundations will be repositories of all humanity's knowledge. They are established under conditions that will allow the forces of history to bring about a second empire in only one thousand years instead of the thirty thousand that would otherwise be needed.

Asimov went on to write a whole series of stories in this future universe, including "Bridle and Saddle," "The Big and the Little," "The Wedge," "The Dead Hand," "The Mule," "Now You See It—," and "—And Now You Don't." All were published in *Astounding Science Fiction* during the 1940s. In the 1950s they were collected together and republished as three novels: *Foundation, Foundation and Empire,* and *Second Foundation.* In the 1980s Asimov returned to the Foundation universe and has since written several more novels in the series. It is without doubt the most successful and popular series of SF novels ever published.

One of the most fascinating "characters" in the original Foundation stories was not a human at all. It was a planet, the planet Trantor. Trantor is the center of the Galactic Empire, the capital of the entire Galaxy. It is also located near the *physical* center of the Galaxy. Our first view of Trantor is through the eyes of Gaal Dornick, a young man from the sticks who is venturing into the big city for the first time. He "had never seen Trantor before. That is, not in real life . . ."

He *had* seen it many times on the hyper-video, and occasionally in tremendous three-dimensional newscasts covering an Imperial Coronation or the opening of a Galactic Council. Even though he had lived all his life on the world of Synnax, which circled a star at the edges of the Blue Drift, he was not cut off from civilization, you see. At that time, no place in the Galaxy was.

There were nearly twenty-five million inhabited planets in the Galaxy then, and not one but owed allegiance to the Empire whose seat was on Trantor.

Dornick arrives at Trantor to work for his hero, the legendary Hari Seldon. He wanted to see Trantor from the spaceship's viewport but was not able to do so. After the ship lands, he goes to his hotel (shadowed by the Secret Police). He still wants to see Trantor with his own eyes, so Dornick decides to take the elevator up the Observation Tower.

He could not see the ground. It was lost in the ever increasing complexities of man-made structures. He could see no horizon other than that of metal against sky, stretching out to almost uniform grayness, and he knew it was so over all the land-surface of the planet. . . .

There was no green to be seen; no green, no soil, no life other than man.

Trantor, the planet at the physical, economic, and political center of the Galaxy, is a planet unique in the Empire. Asimov quotes from the (fictional) *Encyclopedia Galactica*:

. . . located, as it was, toward the central regions of the Galaxy among the most densely populated and industrially advanced worlds of the system, it could scarcely help being the densest and richest clot of humanity the Race had ever seen.

Its urbanization . . . had finally reached the ultimate. All the land surface of Trantor, 75,000,000 square miles in extent, was a single city.

In 1941, when Asimov first started writing the Foundation stories, almost nothing was known about the center of the Galaxy. Radio waves had been detected coming from the central region of the Milky Way. No one, however, had any idea whether they were actually coming from the center or from someplace that lay along Earth's line of sight to that region. Radio astronomy as a scientific discipline would not come into existence until after World War II.

Furthermore, it was impossible to visibly see what was located in the Galactic core, or what might be happening. Dense clouds of opaque gas and dust block all wavelengths of visible light coming from the Galactic central regions. Astronomers could make a good guess at the center's appearance by looking at galaxies similar to ours, such as the Andromeda Galaxy (M31) or the Sombrero Galaxy (NGC 4594). It seemed likely that our Galaxy had a thick bulge of stars at its center and that the density of stars in the very central regions was quite high. Astronomers also knew from visible and telescopic observations that the star clouds in the Milky Way get denser and denser near the center of the Galaxy.

But that was the extent of knowledge about the Galactic center in the 1940s. Asimov himself had read almost nothing about the center of the Galaxy when he began writing the Foundation stories. He did know that it was located in the direction of the constellation Sagittarius and was about thirty thousand light-years from Earth. And that was about it. Neutron stars and black holes were theoretical fantasies in the minds of then-obscure physicists. White dwarf stars were the most bizarre stellar objects then known. When Asimov created the imaginary planet Trantor, he thought it perfectly reasonable to place Trantor near the center of the Galaxy. There was no known reason why a planet inhabited by forty billion people and almost totally covered by a continuous metallic city could not exist there.

Today we know that even if a planet existed near the center of the Galaxy, it could not be at the very center. Nor could it possibly be a place where even four hundred people could live, much less forty billion. But that is now. "Foundation" was then.

The Center of the Galaxy II: The Exploding Core

One of the most popular science fiction writers today is Larry Niven, the master of what is called "hard" science fiction. Hard SF (as opposed to soft SF or to fantasy) is science fiction in which the science is as realistic and as sharp-edged as possible. It is science fiction in which the science and the imaginative extrapolations from science are really the main characters.

Niven's first stories were published in 1965. But he truly burst onto the SF scene in October 1966, with the publication in *Worlds of If* magazine of his award-winning story "Neutron Star." That was followed in rapid succession by "At the Core" (November), "Relic of the Empire" (December), "The Soft Weapon" (February 1967), "Flatlander" (March), and "The Ethics of Madness" (April). Along with one other story, these were later gathered into the short-story collection *Neutron Star*, published in 1968 by Ballantine Books. It has been in print continuously ever since.

In his story "At the Core," Niven paints a vivid picture of a then-current scientific speculation about what might be happening at the center of the Galaxy. The main (human) character is

Beowulf Shaeffer, the protagonist of "Neutron Star" and several
other Niven stories. Shaeffer is hired by the alien puppeteers to
test-fly a spaceship fitted with a new type of faster-than-light
drive. This souped-up space buggy will travel at a velocity of
about one light-year per *minute*. A four-light-year trip would
ordinarily take twelve days with this fictional universe's standard faster-than-light drive. It would take five minutes using the
puppeteers' experimental engines.

A trip to the Galactic core would ordinarily take some three
hundred years, but Shaeffer now realizes it will be only a
month's travel. "No known species had ever *seen* the Core," he
thinks to himself. "You can find libraries of literature on those
central stars, but they all consist of generalities and educated
guesses. . . . Three centuries dropped to less than one month!
There's something anyone can grasp. And with pictures!"
Shaeffer is nothing if not a dyed-in-the-wool capitalist. The
puppeteers offer him a princely sum to test-fly the new ship to
the Core, and he quickly accepts.

A serious problem quickly crops up, however, supplying
spice to the story. The spaceship moves *so* fast that Shaeffer
continually has to dodge around stars. The stress becomes too
much, and he frequently has to drop out of hyperspace to rest.
His contract with the puppeteers requires him to reach the core
and then return within four months. Shaeffer calls his employer
on hyperspace radio to ask for a contract modification, and is
refused. He must pay a financial penalty for each day of delay,
and it will be deducted from his agreed-upon fee of one hundred thousand stars. Grudgingly, desperately, greedily, Shaeffer
pushes on.

Three and a half weeks later, he begins to see the core of
the Galaxy:

> Gone were the obscuring masses of dust and gas. A
> billion years ago they must have been swept up for fuel by
> the hungry, crowded stars. The Core lay before me like a
> great jeweled sphere. . . . A clear ball of multicolored light
> five or six thousand light-years across nestled in the heart of
> the galaxy, sharply bounded by the last of the dust clouds. I
> was ten thousand four hundred light-years from the center.

Shaeffer briefly fantasizes about finding a planet circling one of those stars and staking his claim to it. "Man, what a view!" he thinks.

But not for long.

He soon notices something peculiar—a "blazing, strangely shaped white patch in the Core." He reports to his puppeteer employer, who comes up with an answer. The stars in the core are an average of half a light-year apart, and even closer together at the very center of the Galaxy. They shed enough light on each other to substantially raise each other's temperature and thus speed up their evolution. Therefore, the core should logically have a larger proportion of supernovae than elsewhere. A supernova is the last ultra-explosive stage in the life of an extremely massive star. The puppeteer tells Shaeffer that in the crowded confines of the core, one supernova explosion would likely set off a similar explosion in several other nearby stars. So the white patch that he sees is the light from a small supernova chain reaction that took place ten thousand years earlier. Not to worry; just take a detour around it.

Over the next several hours Shaeffer moves in closer and closer. The light from the core gets brighter and brighter, and is soon intolerable. As the radiation sensors outside the ship chatter a death song, the truth of what has actually occurred in the Galactic core dawns on him:

> Sometime during the morning of the next day I stopped. There was no point in going farther.
>
> "Beowulf Shaeffer, have you become attached to the sound of my voice? I have other work than supervising your progress."
>
> "I would like to deliver a lecture on abstract knowledge."
>
> "Surely it can wait until your return."
>
> "The galaxy is exploding."
>
> There was a strange noise. Then: "Repeat, please."
>
> "Have I got your attention?"
>
> "Yes."
>
> "Good. . . ."
>
> "You say the galaxy is exploding?"

"Rather, it finished exploding some nine thousand years ago. . . . A third of the Core is gone already. The patch is spreading at nearly the speed of light. I don't see that anything can stop it until it hits the gas clouds beyond the Core."

The puppeteer is appalled. Like many other galaxies that have been observed telescopically, the Milky Way Galaxy has an exploding core. A spherical wave of intense x and gamma radiation is even now sweeping out from the center of the Galaxy, the puppeteer notes. When the wave of radiation reaches the puppeteer's worlds, it will sterilize them. With a little bit of arm-twisting, Shaeffer convinces the puppeteer to release him from the contract. He took only a month to get within ten thousand light-years of the Core. He returns home at a more leisurely pace (in two months) to discover that the puppeteer alien race has already fled known space. They have disappeared, fleeing the Galaxy in search of safety. It makes no difference to them that the killing wave of radiation will not reach the human- (and puppeteer-) occupied parts of the Galaxy for another twenty thousand years. The puppeteers are natural cowards, and they've left town.

The image of supernova chain reactions, which is the pivot point of "At the Core," was a perfectly legitimate scientific speculation in the late 1960s, when Niven wrote the story. It was well known that many galaxies had cores that appeared to be exploding. Galaxies called Seyfert galaxies, discovered by Carl Seyfert in 1943, were also known to have very bright cores. The question was, what could be the cause of these bright cores, or of the explosions in the centers of these galaxies? The supernova chain reaction was one possible answer that seemed reasonable at that time.

Niven based his description of the Galaxy's core on what he had read about Seyfert and other so-called "peculiar" galaxies. Niven's assertion that *our* Galaxy has an exploding core was purely fictional; it's the entire raison d'être for "At the Core." His description of the mechanism for such a stellar chain reaction, however, was a quite logical one. So is the image of an expanding wave of killing radiation flooding out from such a core explosion.

Two postscripts to Niven's story "At the Core": First, he wrote it in 1965, and it was published in 1966. The x-rays and gamma rays coming from the Galactic center were not detected until 1970 and 1971. And it has only been in the last ten years that astronomers have begun to gauge the true intensity of the energy and radiation fields that exist at the center of our Galaxy. Second, several years later, Niven wrote a short story entitled "Down and Out," in which he *did* postulate the existence of a supermassive black hole at the center of the Galaxy. It later became part of the second chapter of his novel *A World Out of Time*. Like any good SF writer, Niven had adroitly used the latest scientific information as a springboard for a story.

So did physicist/SF writer Gregory Benford.

The Center of the Galaxy III: The Eater

Gregory Benford is an internationally renowned physicist and astronomer, a professor at the University of California, Irvine, and the author of more than a hundred scientific papers. They include highly technical studies of processes that may be occurring at the center of the Galaxy. Benford is also an award-winning science fiction writer, whose books include *Timescape*, *Artifact*, *In the Ocean of Night*, *Across the Sea of Suns*, and *Great Sky River*. The last three are set in the same future, a future in which humanity has ventured into the regions near the Galactic center, has achieved astonishing heights of civilization, and has then been reduced to small bands of people hiding from incomprehensible machine life-forms.

Benford's 1989 novel, *Tides of Light*, continues the story begun in *Great Sky River*. The main character is Killeen, who now leads Family Bishop on a journey to a new home. In an ancient spaceship named *Argo*, they have fled the world Snowglade, which circles a sun near the Galactic center. Killeen and his tribe will eventually meet other humans, form a strange alliance with a machine life-form, and make contact not only with Killeen's dead father, but with a form of life even stranger than the machine-life.

The vision of the Galactic center that Benford paints in *Tides of Light* is nothing like that of Asimov's Foundation universe. It also differs from Niven's vision in "At the Core." The difference is mainly caused by time and increasing knowledge about the conditions that may actually exist at the center of the Galaxy. In the opening pages of the novel, we see the Galactic center through the eyes of silver-space-suited Killeen as he walks the hull of *Argo*:

> His first impression, as he raised his head to let in all the sweep of light around him, was of a seething, cloud-shrouded sky. He knew this was an illusion, that this was no planetary sky at all, and that the burnished hull of the Argo was no horizon.
>
> But the human mind persisted in the patterns learned as a child. The glowing washes of blue and pink, ivory and burnt orange, were not clouds in any normal sense. Their phosphorescence came from entire suns they had engulfed. They were not water vapor, but motley swarms of jostling atoms. They spilled forth light because they were being intolerably stimulated by the stars they blanketed.
>
> And no sky back on Snowglade had ever crackled with the trapped energy that flashed fitfully between these clouds. Killeen watched a sprinkle of bluehot light near a large, orange blob. . . . It coiled, clotted scintillant ridges working with snakelike torpor, and then burst into luridly tortured fragments.

In Benford's fictional Galactic center, space is clotted with ropes and clouds of dust and gas. Immense fluxes of energy and radiation cause the gas and dust to flare into incandescence. The sky roils with energy. As gargantuan lightning bolts flash back and forth, Killeen gets glimpses of the source of all the energy:

> Anger forked through the sky. Behind them spun the crimson disk of the Eater, a great gnawing mouth. It ate suns whole and belched hot gas. In *Argo*'s flight from Snowglade, which swam near the Eater, they had beaten out against streaming, infalling dust that fed the monster. Its great disk was like burnt sugar at the rim, reddening steadily toward

the center. Closer in swirled crisp yellow, and nearer still a bluewhite ferocity lived, an enduring fireball.

What Killeen and his people call "the Eater" is a black hole with the mass of a million suns. It "eats suns whole." Surrounding the Eater is what physicists call an accretion disk. It is a disk of gas and plasma fed by the surrounding matter, which is pulled in by the black hole's immense gravitational field. It glows red and yellow with the heat and other energy generated as it rapidly rotates around the black hole. The black hole itself is invisible, but Killeen sees the "bluewhite ferocity" of light caused by matter spiraling into it. It is an awesome vision for Killeen, a nonscientific man thrust into an environment beyond his comprehension:

Here tides of light swept the sky, as though some god had chosen Center as her final incandescent artwork.

Benford is using the latest scientific knowledge of the Galactic center in these descriptions from *Tides of Light*. And as we ourselves will see during our own journey to the center of the Galaxy, at the core there is no planet Trantor, no Foundation of a new Galactic civilization. As best we can tell, Benford is correct. The center of our Galaxy is awash in tides of light.

And at the very center is the Eater of suns.

PART II
A JOURNEY THROUGH TIME

2
MYTHS OF THE CENTER
AND THE COSMOS

The contemporary definition of cosmology is "a branch of astronomy that deals with the origin, structure, and space-time relationships of the universe." (*Webster's Ninth New Collegiate Dictionary*). For prescientific cultures, however, a cosmology is a myth that serves as both a model and a likeness of the sacred and physical order of the cosmos. This is an important distinction, and we have to keep it in mind as we examine stories from different ancient societies that identify the center of the cosmos and tell of the creation of the universe and the origin of the Milky Way.

Myths and Motifs

Myths are not lies, or meaningless fairy tales, or crude attempts by primitive cultures to explain the universe. They are all-embracing models for the order and meaning of life and reality. Joseph Campbell, one of our century's great explainers of myth, has said that a mythology has four main functions: mystical, cosmological, sociological, and pedagogical. The first function of a mythology is to awaken and sustain in a person the sense of wonder about and participation in this mysterious universe. A mythology's second function is to fill every part of the current cosmology with this sense of awe and mysticism. Its third function, says Campbell, is to endorse and support the current sociological system and its structure of morals and ethics. Its fourth function is to guide a person through the transitions of his or her life: birth, childhood, sexual awakening, adulthood, parenthood, old age, and finally death.

Nearly every culture past and present that we have discovered and studied has some kind of guiding mythology, filling these four major roles in the lives of individuals and societies. The one significant exception, in Campbell's view, is us. Our contemporary culture, he has said on many occasions, is a "terminal moraine" or garbage heap made from the fragments of many different mythologies. The overarching mythology was once that of Christianity. That was replaced by the Newtonian worldview of the seventeenth through the nineteenth centuries. Now even that is shattered.

Today the sciences of quantum mechanics, relativity, and ecology do give us a sense of awe and even mysticism about the cosmos. That sense does indeed permeate current cosmologies of the expanding universe, many-universe visions of quantum physics, and earth as Gaia, a living entity. However, science's role as the supporter and enforcer of the current sociological system has broken down. And it never really had much of a function as a guide for people through their major life transitions. Today we search for meaning in a universe that seems to have nothing but built-in uncertainty. We search for a center in a cosmos that seems to have none.

A frog will not see a fly unless the fly moves. Its brain is hard-wired to perceive only moving objects and to interpret as "food" only small moving objects. This same kind of selective vision is true of all creatures, including humans. Our perceptual filters are not only of the "hard-wired" kind, but also of the "software" or learned type—we cannot see in the infrared, for example, but smells can trigger memories long forgotten. When we read the creation stories and cosmology myths of prescientific cultures and societies, we inevitably and unconsciously interpret them in our own contemporary terms. We "see" them not with their eyes, but ours. We filter their original meaning through our current cultural biases and our scientific worldview.

It's important to remember that the cultures of the Babylonians, Hebrews, Greeks, Aztecs, Chinese, and other ancient peoples were not permeated with our current scientific knowledge. They looked at the night sky, day after week, month after year after century after millennium, and *knew* that all revolved around the Earth. They *experienced*, in a real and concrete way, the immovability and solidity of the Earth. Terra was quite firma, and was obviously the physical center of the physical universe.

However, the worldview of ancient peoples was much broader than this. The realm of the gods and spirits was every bit as real—if not more so—than the physical world. The two were also intimately connected. To speak of the Earth as the center of the cosmos was to make more than a statement of physical cosmology; it was also a statement about a spiritual reality. The Milky Way was not a partial view of our Galaxy of a trillion stars. It was a "milky way," a stream of divine milk, or the home of gods, or a river in the sky with spiritual significance. The "center of the world" or of the universe was not just a physical location; it was a significant and sacred place.

Western society of the late twentieth century has accumulated immense amounts of information about physical space and physical time. Einstein's two theories of relativity, special and general, and the theory of quantum mechanics are the culmination of that kind of knowledge. However, our highly scientific and technological culture has almost no knowledge (either intellectual or spiritual) of *sacred space and sacred time*. We make little room for sacred space and time in our lives. But sacred space and time were of profound importance to our ancestors.

Nearly all the ancient cosmologies have several features in common. First, the Earth is seen as having a center, a mysterious place that connects the Earth and its inhabitants with the cosmos. It is often represented as a mountain, a tree, or some sort of axis that unites heaven and Earth. For example, many of the peoples who lived in the northern regions of our planet saw the center as an axis that stretched to the polestar.

Further to the south, other cultures identified the center of the Earth with a holy mountain. For the Indians it was Mount Meru; for the Sumerians, Mount Sumer. The Hebrews identified the Earth's sacred center with Mount Zion. The Muslims saw it as the Holy Rock of Jerusalem, upon which was built the Dome of the Rock mosque. At the same time, the Kaaba in Mecca, with its mysterious black rock, might also be considered a "sacred mountain" in the Islamic cosmology. The ancient Egyptians identified the sacred center with the Mountain of the Moon, the legendary source of the life-giving Nile River. These and other cultures often build replicas of the central sacred mountain— Egyptian pyramids, Babylonian ziggurats, the stepped temples of the Aztecs, and Silsbury Hill in England are all examples.

The Earth's sacred center is also represented by the motif of the World Tree. This powerful symbol appears in many religions and cosmologies, from the ancient Chinese to Native Americans, from Siberian shaman costumes to the Tree of Life in the Hebrew scriptures. Yggdrasil, the great ash World Tree of Norse mythology, is a wonderful example of this motif.

Of course, these prescientific peoples also considered the Earth to be the *physical* center of the cosmos. That was patently obvious to anyone who spent any time watching the way the sun, moon, and stars moved about the Earth. However, this physical reality was nowhere near as important as the spiritual reality represented by the Sacred Mountain or the World Tree.

Another common aspect of these cosmologies is the importance of directions. Each of the four directions (north, east, south, and west) has spiritual and cosmological significance and is often represented by different colors, plant and animal spirits, and powers. In these cosmologies there is a harmony among all the different parts of the universe. The worlds of the gods and spirits, and a place or realm of existence called "paradise" (and sometimes a dark counterpart), are integral parts of the cosmos and are included in the cosmology.

Finally, the universe is often depicted as having a kind of layered structure. Frequently humans must move through the different layers in a journey from darkness to light. The Hebrew rabbinical tradition, for example, uses features of Babylonian cosmological images for its picture of the cosmos as a "seven-storey mountain." The underworld (known as Sheol) also has seven levels, pointing downward like an inverted seven-stepped pyramid. This image of double-pyramidal mountains joined at their bases is actually a common one in many cosmologies. The ancient Chinese believed in the existence of an "inner Earth" within the one we know, a mirror image of the world above. Indian mythology speaks of seven heavens; Buddhist cosmology has (variously) seven, nine, or twenty-seven heavens and eighteen hells. The Aztec cosmology has thirteen layers of heaven and nine underworld levels.

In addition to these basic features, early cosmologies often assigned different meanings or roles to various plants and animals, to terrestrial phenomena such as the wind, storms, lightning, and clouds, and to astronomical objects such as the sun, the moon, the stars, the planets—and the band of light that flows across the night sky, the Milky Way.

Myths of Creation and the Milky Way

The myths most familiar to Westerners are those of the Greeks and Romans, our cultural ancestors. It is from these myths that the name "Milky Way" comes. According to Greek myth, the famous hero Heracles (also called Hercules) was the son of Zeus and his lover Alcmene. Hera, Zeus's wife, was naturally resentful of this bastard son of her husband. Heracles felt much the same way about her. Once, while still a very small child, Heracles bit Hera's breast. Some of Hera's milk spilled out and formed the Milky Way—thus, the name it still has today among the philosophical descendants of the Greeks.

The Pima are a tribe of Native Americans living in southwestern Arizona. They are probably the descendants of a mysterious native community called the Hohokam. The Hohokam in turn were the descendants of people who migrated north from Mexico. The Pima are therefore likely descended from the people who later became the Mayans and Aztecs. The Pima creation story presents the first man as the creator of the Earth, the stars, and the Milky Way. The story is recounted by Natalie Curtis in her 1907 book *The Indians' Book* and by Frank Russell in "The Pima Indians," part of a U.S. government report from 1908, and reproduced by Joseph Campbell in his classic *The Way of the Animal Powers*:

In the beginning there was only darkness everywhere—darkness and water. And the darkness gathered thick in places, crowding together and separating, crowding and separating until at last out of one of the places where the darkness had crowded there came forth a man. . . .

[The man] put his hand over his heart and drew forth a stick. . . . Then he made for himself little ants; he brought them from his body and put them on the stick. . . . The stick was of greasewood, and of the gum of the wood the ants made a round ball upon the stick. Then the man took the ball from the stick and put it down in the darkness under his foot, and as he stood upon the ball he rolled it under his foot and sang:

> I make the world, and lo!
> The world is finished.
> Thus I make the world, and lo!
> The world is finished.

So he sang, calling himself the maker of the world. He sang slowly, and all the while the ball grew larger as he rolled it, till at the end of his song, behold, it was the world. . . .

Now the man brought forth from himself a rock and divided it into little pieces. Of these he made stars, and put them in the sky to light the darkness. But the stars were not bright enough.

So he made Tau-muk, the Milky Way. Yet Tau-muk was not bright enough. Then he made the moon. . . .

One of the major figures in the Piman creation myth is Siuhu, or Elder Brother. The Piman myth is one of a class of creation myth common to the American Southwest called the Emergence Myth. The core of this myth is the image of the human race coming from within the earth. In the Piman story, Siuhu emerges from the center of the world, then returns to lead his people from beneath the earth. As he returns to the underground to lead forth the people, Siuhu sings:

> Here I have come to the center of the Earth;
> Here I have come to the center of the Earth.
>   I see the central mountain;
>   I see the central mountain.

The sacred central mountain is a familiar motif in the mythologies of many Native American tribes, including the Pima and their cousins, the Aztecs. The famous stepped pyramids of the Aztecs, today the focus of pilgrimages by thousands of American tourists, were originally symbols of a sacred central mountain in Aztec mythology. The connection between the sacred cosmos and the physical cosmos also is present. The Aztec pyramids, like those of the Egyptians, are carefully oriented so their four sides face the four prime directions.

Sacred mountains also figure prominently in many other prescientific cosmologies. Perhaps surprisingly to some, one is the cosmology of the ancient Hebrews of the Near East. Mount Zion, upon which was built the Holy City of Jerusalem, is the center of the universe in this cosmological vision. Surrounding the Holy Mountain is the land of Israel. Beyond that are the waters of the ocean and the lands beyond. Rising from those farther shores is the vault of heaven.

One of the most powerful renditions of the "sacred center" motif in myth is found in the vision of Black Elk, the Oglala Sioux shaman. Black Elk's vision is recounted by poet John Neihardt in his book *Black Elk Speaks*. Black Elk was a veteran of Custer's Last Stand, a survivor of the massacre at Wounded Knee, and one of the last surviving Sioux shamans from the days before the domination of the white man. Black Elk was about eighty-one years old when he recounted his vision to Neihardt in 1944. Black Elk's vision, which he apparently experienced when he was only about nine years old, also incorporates most of the major motifs of a prescientific cosmology. Present in his vision are the Central Mountain, the World Tree, the Four Directions, the unity of all beings and all creation, the unity of the sacred cosmos with the physical cosmos, and even a hint of a layered reality. Said Black Elk:

> I was on my bay horse, and once more I felt the riders of the west, the north, the east, and the south, behind me in formation, and we were going east. Then I was standing on the highest mountain of them all, and round about beneath me was the whole hoop of the world. And while I stood

there I saw more than I can tell, and I understood more than
I saw; for I was seeing in a sacred manner the shapes of all
things in the spirit, and the shape of all shapes as they must
live together like one being. And I saw the sacred hoop of
my people was one of many hoops that made one circle,
wide as daylight and as starlight, and in the center grew one
mighty flowering tree to shelter all the children of one
mother and one father. And I saw that it was holy.

Black Elk then told Neihardt that the mountain upon which
he and his bay horse stood was a real place—Harney Peak in the
Black Hills of South Dakota. He added, "But anywhere is the
center of the world."

There is a remarkable and not-insignificant resonance between this comment by Black Elk and one first written twenty-
five hundred years earlier. *The Book of the Twenty-Four Philosophers* is a twelfth-century Latin translation of sacred writings
from the Hellenistic period of the fifth century B.C. It contains a
statement that has been quoted over the centuries by Western
philosophers ranging from Nicholas of Cusa to Voltaire. Today,
in the "terminal moraine" of the late twentieth century, with its
vision of a centerless expanding universe, it seems to have more
meaning than ever. "God," said the unknown Hellenistic writer,
"is an intelligible sphere whose center is everywhere and whose
circumference is nowhere."

The Piman people perceived the Milky Way as a creation to
light the night. Halfway around the world, the Milky Way is also
visible in Australia's night sky. Australian aborigines see the
Milky Way not as a river of light, but as the smoke from a
heavenly campfire. The fire was lit by Nagacork, the One who
created the world, after he finished his work and went to sleep
among the stars. For a people who once lived much of their lives
in the open, cooking food and keeping warm with a smoking
campfire, this vision of the Milky Way makes perfect sense.

The same is true of the Milky Way explanation of the Turko-
Tatars of central Asia. These people lived in tents, and to them
the sky was an even greater tent. The stars were holes in the tent.
The polestar was the top of the tent pole rising up from the

north and holding the sky-tent aloft. The Milky Way, finally, was the seam of the sky-tent.

Further to the north were the Chukchi of northeastern Siberia. They saw the polestar as the axis of the world between morning and evening. The axis was linked to morning by the Milky Way.

Somewhat different, and more complex, is a Chinese myth that depicts the Milky Way as a river in the sky. The heavenly river plays a part in the story of the Herdsman and the Weaver Girl. The Weaver Girl (which is also the star we know as Vega) was a shy young woman concerned with her work around the home of her father, the One Who Stands Outside Time. Her father arranged for Weaver Girl to marry the Herdsman (which is also the constellation roughly equivalent to our Aquila, the Eagle). The Herdsman tended cattle on the far side of the Milky Way. He came over to the near side, married the Weaver Girl, and they were very happy together. However, the Weaver Girl became so lost in her love for her husband that she began to neglect her work around her father's house.

That disturbed the father, so he arranged to change things. He called together a flock of magpies, who made a winged bridge over the river of the Milky Way. Then the father ordered the Herdsman to cross the bird bridge to the far side and resume herding the cattle. The Herdsman did so, and the magpies flew away as soon as he reached the far side of the Milky Way. The Weaver Girl returned to her homely duties, and the Herdsman resumed his herding in that dark land which is hidden from human eyes. Once a year, on the seventh day of the seventh month, it is possible for the two lovers to briefly visit together. People pray that the weather will be clear that night. Then, it is said, the magpies will re-form the winged bridge over the Milky Way, and the Weaver Girl can cross over the silver sky river to visit her husband the Herdsman.

Cosmologies and Cosmograms

The Desana are a tribe of native South Americans, living in an area of Colombia near the border of Brazil and Venezuela. Their land lies very near the Earth's equator. The cosmology of the

Desana people includes several of the motifs common to most prescientific cosmologies. The Desana myths have been transcribed and interpreted by Gerardo Reichel-Doimatoff, an anthropologist at UCLA. The cosmos, according to the Desana, was created by the pure will or "Yellow Intention" of the Sun Father. The Creator is not the same as the sun, but rather a being of eternal yellow light. The rays of the Creator's Yellow Intention continue to permeate all of reality.

The Desana cosmology depicts a universe that has three levels. The Earth is the middle level and is symbolized by the color red. Below the Earth is a River of Milk. It is represented by the color green, the color of coca leaves. The upper level has two zones. One is the realm of the sun and moon, and has the colors orange, yellow, and white. The other zone is the Milky Way, which has the symbolic color blue. The Milky Way arises from the River of Milk.

The Desana cosmology sees the Milky Way as a huge, dangerous river, arching over the Earth and moving from east to west. The Milky Way contains the seeds of life, but it is also filled with disease and death. As it flows across the heavens, the Milky Way is tossed and turned by the Wind Current. This zone of the heavens, the zone of the Milky Way and the Wind Current, is the region to which people go when they are under the influence of the hallucinogenic drug *vihó*.

The guardian of this drug is a figure called Vihó Man. Vihó Man lives in the Milky Way in a state of eternal trance. The shamans or medicine men of the Desana speak with the Vihó Man in order to enter into the mystical trance by which they come into contact with other invisible beings of the Desana religion, such as the two Animal Masters, or the Jaguar. Thus, for the Desana, the Milky Way is much more than a band of light in the night sky or even a dangerous cosmic river of life and death. It is the place where humans and supernatural beings meet and communicate with one another. To the far west of the Earth is a region of darkness, home of sorcerers and witches who punish those who disregard the tribal laws and taboos.

The creation myth of the Desana states that the Sun Father determined the center of their homeland by finding a place where His stick would cast no shadow. By placing His stick at this spot, the Sun Father impregnated the Earth. The opening became a whirlpool leading into and out of the Earth's womb. The Creator then produced an ancestral male for each of the many tribes of the region (including the Desana), and they emerged from the hole in the ground made by the Creator's stick. A spiritual guide called Fermentation Man or Flowing Forth Man accompanied each ancestor as he sailed out of the whirlpool entrance on canoes made of living anacondas. The canoes took each tribal ancestor to the land where that tribe would live.

The location of the world's center in the Desana cosmology is in sacred space, of course. But it is a physical place as well. In a river that runs not far from the Desana land, in the territory of Colombia's Barrasana tribe, is a huge boulder called the Rock of Nyí. Carved into the Rock of Nyí are a series of petroglyphs that depict the creation story of the Barrasana and their Desana neighbors. It supposedly marks the spot where the Sun Father planted his stick and it cast no shadow. In fact, this location happens to be almost exactly on the Earth's equator. On the spring and autumn equinoxes, the sun rises due east and passes almost directly overhead. At noon on the equinox, a straight stick planted upright in the ground in this location *casts no shadow*. In a spiritual or mythological sense, the sun's rays are penetrating directly into the Earth, just as the Sun Father's energy did at the creation of the center of the world, and of humanity.

So the Desana cosmology depicts a cosmos with different levels. Humans entered the world by climbing up to the middle level (the Earth) from a lower level (the River of Milk). In the upper realm live supernatural creatures like the Vihó Man and other invisible beings, with whom humans communicate through the intercession of the shaman. The cosmological picture includes a "World Mountain"—the sacred Rock of Nyí. It also has a kind of "World Tree"—the Creator's shadowless stick and the vertical rays of the sun on the equinoxes symbolize the unity of the heavens, the Earth, and the underworld.

Quartz crystals play an important role in the Desana's

shamanistic spirituality. Along with the hallucinogenic powder made from the *vihó* vine, shamans use quartz crystals in their rituals. The crystals are often referred to as crystallized semen. The Desana worldview mirrors their importance. The cosmos has a six-sided structure, both physically and spiritually. For example, the Desana believe that the borders of their territory were established by six anacondas stretched out to form the six sides of a hexagon. A giant hexagon of stars centered on Orion's belt is symbolically linked to the six waterfalls that mark Desana territory. The center of the celestial hexagon is connected to the location on the Pira-Pirana River, where the Sky Father planted his shadowless stick. Thus, heaven and Earth are intimately linked into one whole—which, of course, is one of the major functions of a cosmology.

The images or visions of the cosmos and its center can be given graphic form in a **cosmogram**. The Desana cosmogram, for example, would show a flattened disklike Earth floating in the void. Beneath it would be an oval "River of Milk." Rising out of the River of Milk and flowing up and over the Earth from east to west would be the Milky Way. Above and beyond the Milky Way would be the sun and moon.

The cosmogram of another South American tribe, the Warao of Venezuela, is somewhat different in shape but similar in other details. The Earth is a flat disk floating in a world sea. A cosmic four-headed snake, the Goddess of the Nadir, encircles the Earth in the Underworld realm. Each of its heads is at a cardinal direction (north, east, south, west). Another snake, the Snake of Being, lives in the world sea and also encircles the Earth. The sky is a bell-shaped tent. Its highest point is supported by the world axis. High in the sky, at a level whose borders are marked by the highest angles of the sun on the summer and winter solstices, is a realm of Warao heaven. This is the location of the "House of Tobacco Smoke," where the Warao shamans travel by climbing tobacco smoke ropes entwined along the world axis.

This cosmogram is based on factual observation, as far as the Warao are concerned. They live near the equator, so the highest point of the sky's dome (the zenith) is much more important and obvious than the celestial north pole lying near the northern horizon. Clearly (to them), they live at the center of the world; they need only look straight up to see that. The sun's daily paths during the summer and winter solstices are nearly symmetrical on opposite sides of the zenith, thus marking the heavenly realm of the House of Tobacco Smoke. At the spring and autumn equinoxes, however, the sun reaches its highest point in the sky—the zenith. That suggests to the Warao that the sky is tent-shaped, with its highest point nippling up as if held by a central tent pole.

One cosmogram with which many people are somewhat familiar is that of Hindu mythology. This is the one with the giant turtle. The turtle is swimming in the cosmic sea. Standing on the turtle's back are four great elephants. Resting on the elephants' backs is the circular disk of the Earth. India, of course, lies at the center of the Earth, and thus of the cosmos.

In his delightful book *A Brief History of Time*, physicist Stephen W. Hawking tells the (probably apocryphal) story of an eminent scientist who gave a public lecture on the current cosmological vision of the universe. The planets circle the sun, and the sun circles the center of the Galaxy, along with the rest of the Galaxy's stars. A little old lady at the back of the room then stood up and said, "That, sir, is complete rubbish. The Earth is a great plate resting on the back of a giant tortoise." "Aha," replied the scientist, "but then what is the *tortoise* standing on, eh?" "You think you're very clever," the little old lady replied triumphantly, "but it's turtles all the way down!"

Peoples like the Desana and Warao may appear to us Westerners as primitives, living with hardly any clothing and a woeful dearth of technology in the middle of the South American jungles. The truth, however, is that they are sophisticated peoples in their own right, with cosmologies that are effective, functional, and contain immense power. The cosmologies of South American natives give us a clear look at the functioning of the prescientific mythologies of now-vanished cultures and peoples. But they are not the taproot of our mythology. Our contemporary vision of the cosmos and its center—and the Milky Way—has its roots in the Middle East and a peninsula in the eastern Mediterranean Sea now called Greece. But before Greece there was Babylonia. And the peoples and cosmology of the Fertile Crescent have had their influence on us, too.

Marduk vs. Tiamat, and the Babylonian Cosmology

Beginning more than six thousand years ago, a series of cultures rose and fell in an area of the Middle East centered around the Tigris and Euphrates rivers. These lands are now known as Iran, Iraq, and Syria. The Babylonian, Assyrian, and Mesopotamian civilizations—and their cosmologies—dominated the region for centuries.

The Babylonian creation myth centers on the struggle between two great deities, Tiamat and Marduk. Tiamat is female, fruitful, spontaneous in nature. Like the goddesses Ishtar and Innana of other Middle Eastern cultures, she is an avatar or embodiment of the Great Mother Goddess, who was the first identifiable deity of humanity. Marduk, by contrast, is male, dynamic, aggressive, proud. Like Zeus and Yahweh, he is an embodiment of the supreme male god of the tribes who swept into the Middle East from the north and east sometime around 4000–3000 B.C.

In the specific myth of the early Mesopotamian people, Marduk and Tiamat battled for supremacy, and Marduk won. He sliced Taimat in half. He set her bottom half below, where it became the Earth with its oceans and lakes. Tiamat's spotted top half was set on high, and it became the heavens.

More than thirty-five hundred years ago, the Babylonians began placing great emphasis on astronomy and the role of numbers and numerology. By 1600 B.C. they had compiled their first star catalogs. Their interest in the sky also led them to the first historical identification of the five so-called classical planets (Mercury, Venus, Mars, Jupiter, and Saturn) as objects that were different in some ways from the fixed stars. They named those planets for gods and goddesses—indeed, they actually *identified* them with gods and goddesses, which the later Greeks never quite did—and carefully plotted their movements through the sky.

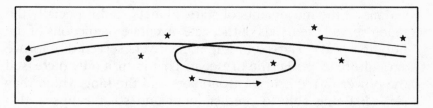

The occasional westerly motion of planets against the fixed background of stars is
called retrograde (or backward) motion.
Illustration from Robert T. Dixon, *Dynamic Astronomy,* fifth edition, 1989. Reprinted by
permission of Prentice-Hall, Inc., Englewood Cliffs, New Jersey.

It was Babylonian astrologers who discovered and mapped
the puzzling retrograde movements of the planets in the heavens. Planets like Mars, Jupiter, and Saturn periodically appear to
reverse their regular counterclockwise motion across the night
sky, and temporarily move "backward." Then the planet will
stop and again move forward in the proper manner. This apparent reversal of movement is called **retrograde motion**. Today
we know that planetary retrograde motion is an illusion caused
by the motion of our own planet around the sun. Prescientific
peoples like the Babylonians and the Greeks did not know this,
however, and came up with involved explanations for retrograde movement.

The Babylonians developed a numerical system based on
12. They identified and named twelve constellations in the
zodiac and developed the first coherent system of astrology. By
800 B.C. they had detailed records of the movements of the
planets through the signs of the zodiac. The Babylonians developed a calendar with twelve months. They divided a circle into
360 parts (12 times 30). Not surprisingly, they had a base 12
counting system. (By contrast, we use a base 10 counting system, with digits running from 0 through 9. Digital computers
use a base 2 system, with digits 0 and 1.) Numbers were mystical entities for the Babylonians.

The Babylonians became careful and accurate astronomers
for a very practical reason. Like the ancient Egyptians, who used
careful astronomical observations to predict the life-giving
flooding of the Nile River, the astronomical work of Babylonian
priest-astronomers was supported by the state. Accurate obser-

vations of the movements of stars, planets, and especially the moon served the needs of the state. Accurate predictions of the flooding of the Tigris and Euphrates rivers served to improve agriculture in the Fertile Crescent. That in turn only increased the power of the priest-astronomers and the king whom they served. Being able to predict lunar and solar eclipses also helped.

The cosmological worldview of the Babylonians, Assyrians, and Mesopotamians influenced the cosmologies and myths of other cultures. The Greek philosopher Pythagoras, for example, is said to have spent many years in the Middle East studying the cosmology of the Mesopotamians. He was greatly affected by Mesopotamian number mysticism, and he incorporated much of it into his own cosmology.

Hebrew religion and mythology also was deeply influenced by that of the Assyrians and Babylonians. A large percentage of the Hebrew people were held captive by the Babylonians (this is the famous "exile" of the Old Testament) for several generations. It's not surprising, then, that large portions of the more mythological parts of Hebrew scripture are borrowed directly or indirectly from Babylonian myths and legends. The Great Flood story is only the best known. Even the venerated Hebrew seven-branched candlestick, the menorah, has a Babylonian connection. It derives from the ancient Mesopotamian symbol of the World Tree. The Genesis creation story that includes the Tree of Life and the Tree of the Knowledge of Good and Evil is also clearly influenced by the Babylonian cosmological vision. Other influences can be seen in the mystical significance of certain numbers and in the very image of Yahweh, the Hebrew god. The mythological death struggle between Marduk and Tiamat was historically reenacted for hundreds of years in the lands of Israel and Judea. The priestly followers of Yahweh fought tooth and nail to destroy the popular folk religion of Ishtar (the Mother Goddess) and her male consort Baal (which means "Lord").

The Sacred Mountain motif also appears in the Babylonian cosmology, as in so many others. The Babylonian ziggurats were symbolic representations of the Sacred Mountain that is the center of the Earth and thus the cosmos. The Tower of Babel story in the Hebrew scriptures is an offshoot of this, and the Tower of Babel itself is clearly a sacred ziggurat. The ziggurats were also used as rather effective observatories by the priest-astrologers, who carefully tracked the movements of the moon, the stars, and the planets.

For all their triumphs, however, the Babylonians and their mythological vision of the cosmos did not prevail. Rather, it was a group of nomads from the shores of the Caspian Sea, far to the north and east of the Fertile Crescent, who would reshape humanity's concept of the universe and open the road that leads to the Monster at the center of the Galaxy.

3
THE GREEKS

The cosmological vision of the ancient Greeks influenced Western thought and philosophy for thousands of years. It was finally superseded only in the seventeenth century A.D. by the combined work of Copernicus, Galileo, Kepler, and Newton. For more than two thousand years, however, it reigned supreme.

The Greek cosmological view was, with one significant exception, geocentric—Earth-centered. It used a series of moving celestial spheres to explain the observed movements of the sun, the moon, the stars, and the *planés* or wandering stars, the planets. Celestial objects that did not fit into this regular scheme of things, such as meteors and comets, were considered messengers from the gods.

The Greek Cosmology

The ancestors of the Greeks were probably tribes living on the shore of the Caspian Sea. Sometime between 2000 and 1500 B.C., they began moving west, toward the eastern shores of the Mediterranean Sea. By 1000 B.C. tribal groups now known as the Achaeans, Aolians, Ionians, and Dorians had settled in the region that is now modern Greece. They established independent city-states, which often warred with each other. The descendants of the Dorians eventually became dominant. The Ionians were forced off the peninsula and settled the islands and coastlands of what is now northwestern Turkey. By the end of the ninth century B.C., the Greek peninsula had been more or less politically united under the rule of kings in the city of Athens.

By the eighth century B.C., the Dorians conquered the city of Corinth. By then they could legitimately be called Greeks, since they controlled the entire peninsula. Just as importantly, they took a dramatic step forward—out to sea. They learned how to navigate by the sun and the stars, and how to read the winds and waves. The Phoenicians were then the rulers of the Mediterranean and traders par excellence, but the Greeks also began to make their influence felt. They established colonies that ranged from Spain in the far west to Sicily and Italy, to the eastern coast of Asia Minor (Turkey).

At the end of the seventh century B.C., Athenian rule was replaced by that of the city-state of Sparta, but it was later reestablished. It was at this time that the Greeks made a conceptual leap forward in their view of the cosmos. Before the sixth century B.C., the Greek mythological cosmology was in most respects similar to that of other prescientific cosmologies. The classic Greek religion of gods and goddesses—Zeus, Hera, Poseidon, Ares, Artemis, Demeter, Aphrodite—had long replaced the more ancient Mother Goddess religion that had held sway for more than thirty thousand years. The cosmology of the Greeks at this time is preserved in its essence in the epic poems of Homer, the *Iliad* and the *Odyssey*, and in the poems of Hesiod. Both probably lived in the eighth century B.C., so it's reasonable to consider their versions of Greek cosmology as somewhat typical.

In this cosmological vision of the universe, the Earth is a flat, circular disk surrounded by a world ocean. The land of the dead, Erebus, is reached by traveling west through the Pillars of Hercules—the Rock of Gibraltar at the western opening of the Mediterranean and the Atlantic Ocean. The heavenly realm lies above the Earth, and its underworld counterpart, Tartarus, lies below. The sky is a roof above the Earth, supported by great pillars.

In the beginning, according to the Greek creation myth, there was only the empty space known as Chaos. Then Gaia (who is a transmutation of the ancient Mother Goddess) or Earth spontaneously appeared. She was joined by Tartarus and by Eros, the creative force embodied in sexual desire and love. From Chaos came Night and Erebus, the mirror of night, which lies under the Earth and separates Earth from Tartarus. Night and Erebus conceived Day and Space, which is the ether of the upper realms. From Gaia came the Sky, which the Greeks called Uranus, as well as the sea and the mountains. Gaia and her son/husband Uranus then conceived the Twelve Titans, the first stewards of the universe. Kronos, one of the Titans, castrated Uranus and took his throne. He later mated with his sister Rhea and became father to the beings who later became the Olympian gods. One of them, Zeus, eventually overthrew Kronos and the other Titans, imprisoned them forever in Tartarus, and with his brothers and sisters took over rule of the cosmos.

In this classic Greek cosmology, Mount Olympus in northern Greece may have functioned as the Central or World Mountain; it was traditionally the home of the Olympian gods. Of course, in a civilization of city-states, each city probably considered itself to be the "center of the world."

If there were any other single spot in Greece that might have that appellation, it was probably Delphi. It was located at the foot of Mount Parnassus in central Greece, a mountain sacred to the gods Apollo and Dionysus, and to the Muses. It was the seat of the Delphic Oracle, without doubt the most famous and powerful spiritual oracle of ancient Greece. Delphi was originally a site sacred to the Great Goddess, and the oracle may have been there long before the classical Greek religion replaced Goddess worship. Delphi then became sacred to Apollo, the sun god. A major shrine was built at the site in the sixth century B.C., and it remained an active oracular site until the beginning of the Christian era.

The Changes Begin

Around the sixth century B.C., a significant change started taking place in the Greek cosmological worldview. It began with several philosophers who lived in the city of Miletus, located in the province of Ionia on the western coast of present-day Turkey. Thales (c. 636–c. 546 B.C.) is the first recorded Western philosopher. He was also, as far as we know, the first to abandon the concept of cosmology as myth, and instead try to explain the origin and nature of the cosmos in purely physical terms.

His cosmology was not so different from that of Homer. The Earth in his view was a flat disk floating in and surrounded by a cosmic ocean. Above the world sea and the disk of the Earth, the sky arched over all. The really significant difference in Thales's view of the cosmos was his belief that the universe had some kind of real, physical origin. That origin was a fundamental element, which Thales said was water. Though he made no estimate of the actual dimensions of the Earth or sky, Thales did believe that the Earth was finite in its dimensions. He supposedly also introduced geometry to Greece and was said to have predicted a solar eclipse that took place in 585 B.C.

Thales's philosophical descendants included Anaximander (c. 611–c. 547 B.C.). Anaximander's cosmological picture was considerably more detailed than that of his teacher Thales. The Earth, which (of course) lay at the center of the universe, was a disk whose diameter was three times its thickness. Surrounding the Earth were several moving rings, like wheels. The sun, moon, planets, and stars were not real, but rather holes in the rings that allowed light from more distant fiery realms to shine through. Anaximander also thought that the sun was the most distant celestial object (or hole) and that the stars were the nearest to Earth. The motivating power for the cosmos was circular motion.

Like Thales, Anaximander suggested that the cosmos was born from a primordial element or principle, not from the efforts of the gods. Unlike Thales, however, he did not think that principle was water. In Anaximander's view it was something he called *apeiron.* Just what *apeiron* actually was is uncertain. It seems to have had the connotation of all-pervasiveness and infinity. Anaximander's universe, in other words, was apparently boundless—infinite in time while still finite in physical extent. This may seem paradoxical, but it is not. The surface of a sphere, for example, is boundless but still finite.

Another Ionian philosopher from Miletus to come up with an interesting cosmology was Anaximenes (c. 585–c. 526 B.C.). Anaximenes claimed that the elemental substance of the cosmos was air, not water or *apeiron.* The universe itself was centered upon the flat disk of the Earth. The sun, moon, and planets were also flat disks, which circled the Earth in their perfectly circular cycles. The stars, said Anaximenes, were attached to a rotating crystal sphere.

The Pythagorean Cosmos

The next major change in Greek cosmology was made by one of the greatest—and most controversial—philosophical minds of Western history. Almost nothing is really known about the life of Pythagoras (c. 582–c. 507 B.C.), and none of his writings survive to the present day. What we do know of him and his teachings comes from the writings of his followers, the Pythagoreans. He was apparently born on Samos, an Ionian island off the coast of Turkey. He traveled widely as a young man, and in middle age left Samos for good and settled at Crotona, a Greek colony town in southern Italy. Some stories suggest that Pythagoras wrote the town's first constitution, designed its money, married, and had two children. What is known for sure is that he founded a mystical and philosophical community whose ideas are still extant today. He was a mystical mathematician.

The underlying principle of the universe, Pythagoras taught, was number. The reason for his belief, say some scholars, lay in his childhood as the son of a gem engraver. Pythagoras became fascinated with the regular geometric shapes of crystals. Different crystals have different numbers of sides: quartz is six-sided, for example, and garnet is twelve-sided. The young Pythagoras came to imagine each gem as actually *made* of a different number. Numbers were not abstractions for him; they were real things. (This is not all that unusual; it is perfectly common for some children to mentally picture numbers as living things—2s dancing about with 7s, 4s hopping along like pogo sticks.) Pythagoras soon discovered the mathematical principles underlying music. Certain pleasant-sounding intervals in a musical scale, for example, are based on simple ratios: a fifth has the ratio 3:2; an octave, the ratio 2:1. Pythagoras thus came to believe that number is the fundamental ground of the cosmos.

Pythagoras's cosmology was a combination of numerological mysticism with some of the earlier ideas of Anaximander and Thales. At the center of the universe was the immobile

Earth—and Earth was a sphere. Circling the Earth were the sun, moon, and five planets, each on its perfect circular path. The stars were fixed in a distant crystalline sphere. As each cosmic object moved along its path, it emitted a tone, which varied according to its distance from the center. The tones and their combinations as they moved made up the music of the spheres—a phrase that survives to this day.

One of Pythagoras's followers, Philolaus of Crotona, modified this scheme somewhat. In place of the spherical Earth at the center of the universe was the Central Fire. This was the source of the sun's light and heat. Circling the Central Fire were the Earth, sun, moon, and the five planets. The stars remained in their outer crystal sphere, and so did the music of the spheres. The other modification by Philolaus had number mysticism as its origin. This cosmology had nine objects in it: the Central Fire, the Earth, the moon, the sun, and five planets. Pythagoreans considered 10 to be a perfect number, and this cosmology was 1 short. So Philolaus added a *counter-Earth* to the scheme! Its motion and the geometry of the cosmos always kept the counter-Earth out of sight of those who lived on Earth.

Other philosophers who lived after the time of Pythagoras came up with different visions of the cosmos. Leucippus (fifth century B.C.) thought the Earth was shaped like a drum. Heracleitus (c. 535–c. 475 B.C.) thought it was bowl-shaped and that

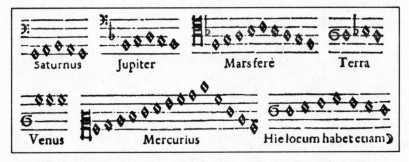

Pythagoras believed that each cosmic object moved along a path fixed to a crystalline sphere, emitting a musical tone as it moved—the music of the spheres. Many centuries later the astronomer Johannes Kepler attempted to write down the cosmic melodies in musical notation.

Illustration from Arthur Berry, *A Short History of Astronomy: From Earliest Times Through the Nineteenth Century,* Dover Publications, Inc., 1961. Reprinted by permission of the publisher.

the fundamental element was fire. Democritus (c. 460–c. 370 B.C.), who came from Thales's school of thought, considered the Earth to be hollow. He also postulated that all things were ultimately made of tiny indivisible entities, which he called *atomos*, or atoms. The constant motion of atoms created the universe, said Democritus. Heavier atoms clustered together to make the Earth, while lighter ones came together to make the heavenly bodies.

A Cosmological Revolutionary

The next philosopher to offer a significantly different cosmology was Anaxagoras (c. 500–c. 428 B.C.). He settled in Athens when he was about twenty years old and is reputed to be the first person to teach philosophy in that famous city. Anaxagoras's cosmological vision of the universe was somewhat controversial in Athens. When he was about thirty-three, a meteorite the size of a wagon fell to Earth in broad daylight near the small town of Aegospotami. Anaxagoras claimed that it had fallen from the sun, which was neither fire nor a god but a glowing stone bigger than all of Greece itself. The stars, too, were glowing stones, only much more distant from Earth than the sun. Their circular orbits, Anaxagoras claimed, took them under the Earth, which was flat and supported by air, not water. The moon, in turn, was a flat disk like the Earth, and shone not from its own light (as many earlier philosophers claimed) but from light reflected from the sun.

Anaxagoras taught that the universe as we know it began when a vortex in the all-surrounding "ether" swung the primordial seeds of the sun, moon, and stars into their regular paths around the Earth. So regular and stable was this vortex that Anaxagoras believed it must emanate from the exact opposite of chaos. The origin of all things, he declared, must be some eternal and perfect Mind.

This rather unusual view of the world—so different in many respects from both the classical mythological vision of gods and goddesses and the musings of Thales, Anaximander, and others—gathered many supporters in Athens. They included

Pericles, the great Athenian statesman, the playwright Euripides, and a young man named Socrates (c. 469–399 B.C.).

Anaxagoras also had his detractors, including the playwright Aristophanes, who made fun of Socrates and Anaxagoras in his play *The Clouds*. Many other Athenians thought Anaxagoras's ideas verged on heresy—a reaction that was repeated many times in the centuries to come. During the Greek wars with Persia, which ran from 490 to 449 B.C., Anaxagoras was arraigned on a charge of impiety. The eloquence of his friend and benefactor Pericles saved him from death, but Anaxagoras was fined and sent into exile in the town of Lampsacus, where he finally died.

Anaxagoras's pupil Socrates, of course, went on to become one of the greatest of all Greek philosophers. He rejected Anaxagoras's cosmology—not because it was too radical, but because it was not radical enough. Socrates objected to Anaxagoras's suggestion that astronomical phenomena had various mechanistic causes. He wanted to take the idea of Mind as origin of the cosmos and, from that one starting point, deduce whether or not the Earth is round or flat, whether it is at the center of the universe or at the edge, and what the ultimate causes of various astronomical phenomena are. Socrates succumbed to the fate that his mentor escaped. In 399 B.C. he was convicted of corrupting the morals of the youth of Athens, and he willingly drank a cup of poison hemlock.

The Platonic Worldview

Socrates himself did not construct a cosmology, as far as we know. It was his most famous pupil, the philosopher Plato (c. 427–347 B.C.), who was the next major cosmological thinker among the Greeks. Plato visited Italy and Sicily after the death of Socrates, later returning to Athens to found his famous academy. He revisited Sicily and the city of Syracuse several times at the behest of its rulers, who were his friends, but always returned to Athens and the academy. Plato was probably exposed to the ideas of Pythagoras during his Sicilian and Italian sojourns. He also came away with considerable knowledge of the goings-on in then-contemporary Italian natural sciences, including astronomy, anatomy, and chemistry. That knowledge all ended up in his famous work *Timaeus*.

Plato's cosmology was influenced by Pythagorean concepts. He believed that the Earth was not flat but a sphere, since the sphere is a perfect shape and thus mirrors the perfection of the cosmos. The sky was not some kind of roof, as Homer had said; it was a sphere. So were the sun, the moon, the planets and stars. All these bodies moved about the Earth in circular orbits. The occasional retrograde motion of the planets could be explained, said Plato, by a complex system of circular movements. Such movements were the visible manifestation of the governing principle of the cosmos. It was not air, or fire, or water; not Mind, or *apeiron*, or number. It was geometry.

Plato included some of his cosmological ideas in his classic books *The Republic, The Statesman*, and *The Phaedrus*. However, in *Timaeus* Plato presented his view of the cosmos as based on his original "Theory of Forms." He envisioned God as a creative force that made all things in the cosmos, using as original patterns certain perfect geometric objects or abstractions. These perfect geometric forms were the five regular solid geometrical forms that we still call Platonic solids. They are the tetrahedron (a solid with four triangular sides), cube (six square sides), octahedron (eight triangular sides), dodecahedron (twelve pentagonal sides), and icosahedron (twenty triangular sides). Thus, Plato's cosmology was as much mystical as rational.

Plato then proceeded to apply his cosmology of geometry and order to politics. This certainly seems a radical move, but in one sense it is not. As with the philosophers and shamans who preceded him, Plato was only trying to put a cosmology to its ultimate use: uniting the realms of heaven and Earth, of the natural and the supernatural, of the physical and the spiritual. The laws that govern the heavens must in some way govern the actions of people on Earth. If such a goal seems strange and "primitive" to scientifically sophisticated people of our times, it is only because we are blind to the functioning of our own guiding worldview. Our contemporary cosmology, knit from strands of astronomy, relativity, and quantum physics, attempts to do exactly the same thing, minus any overt belief in any spiritual or supernatural reality.

The first Greek astronomer to produce an elaborate cosmology based on Plato's suggestions was Eudoxus of Cnidus (c. 400–c. 347 B.C.). Eudoxus was quite a talented man, versed in not only astronomy but also geometry, mathematics, and medicine. Some of his geometric discoveries were later incorporated into the work of his contemporary Euclid. He was also the first Greek astronomer to explain the movements of the heavenly bodies in a manner that might be called scientific.

THE EUDOXIAN SPHERES

Celestial Object(s)	Number of Spheres
Stars	1
Sun	3
Moon	3
Five planets	20 (4 per planet)
Total	27

Eudoxus's description of celestial movements used a series of transparent concentric spheres. An example is his explanation of the movement of the moon around the Earth. Eudoxus used three spheres for this. All three spheres must be thought of as being somehow connected and in motion at the same time. The outermost sphere, said Eudoxus, rotates around the Earth from east to west once every twenty-four hours. Its axis runs through the polestar, which then (and now) is the star Polaris. This sphere's motion explains the fact the moon does indeed seem to move around the Earth once each day, from east to west, along with the stars and planets.

The next sphere inward has its axis tilted at an angle of 23.5 degrees to the outermost sphere. It also rotates from east to west, but with a period of 223 **synodic months**. One synodic month, also known as a lunar month or a lunation, is the interval of 29.53 days between two successive new moons. The

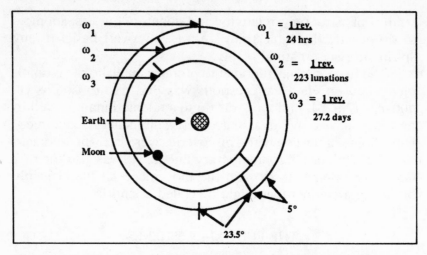

Eudoxus of Cnidus explained the motions of the heavenly bodies by hypothesizing
their attachment to a series of concentric spheres. This diagram shows only the three
spheres that accounted for the moon's motion around the Earth and against the
fixed background of stars. The outermost sphere rotated from east to west once
every twenty-four hours on an axis passing through Polaris, the polestar. The next
sphere rotated from east to west once every 223 lunations (synodic months) and
was tilted at an angle of 23.5 degrees to the outermost sphere. The third sphere,
containing the moon and tilted at an angle of five degrees to the second sphere,
rotated from west to east once every 27.2 days. The complete Eudoxian system
involved twenty-seven heavenly spheres, all somehow connected and all moving
simultaneously.

Illustration from Michael J. Crowe, *Theories of the World from Antiquity to the Copernican*
Revolution, Dover Publications, Inc., 1990. Reprinted by permission of the publisher.

movement of this second sphere explains a slight difference
between two regular lunar periods. The moon's **draconic period** is slightly less than its **sidereal period**. The lunar draconic period of 27.21 days is the time between the moon's
successive upward crossings of the ecliptic, the circle that marks
the yearly path of the sun across the sky. (Astronomers refer to
this as crossing through the **ascending node** of the moon's
orbit.) The moon's sidereal period—a sidereal month—is the
27.32 days the moon takes to complete one orbit around the
Earth, as measured against the fixed background of stars. The
difference of 0.11 days, or two hours and thirty-eight minutes,
adds up over the months and years. The period of 223 synodic
months is a bit more than eighteen years. At the end of this

period, the moon is back in the exact same location in the sky as at the beginning of the cycle.

The innermost sphere is the location of the moon itself, exactly ninety degrees down from the axis of rotation. This sphere's axis is inclined at five degrees to that of the second sphere. It rotates from west to east at a rate of once every 27.2 days. The five-degree angle explains the fact that the moon crosses the ecliptic. It also accounts for the moon's monthly journey through the signs of the zodiac. These are the twelve constellations that occupy a band around the celestial sphere extending about nine degrees on each side of the ecliptic.

With all three spheres in motion at once and all connected to each other, the monthly motions of the moon through its synodic, draconic, and sidereal periods can be effectively explained. So can the moon's movement through the zodiac. This model even explains the longer eighteen-year lunar period, known to cultures around the world for millennia and called by some the "Great Month."

Eudoxus's theory of planetary motions was similar to this but somewhat more complex. The annual movement of the fixed stars was explained by one sphere. Three spheres accounted for the movement of the sun. Eudoxus used four spheres instead of three to account for the motion of each of the five known planets. The fourth was needed to explain planetary retrogression. That added up to a total of twenty-seven revolving spheres to describe the movements of all heavenly bodies (except, of course, meteors and *kometes*, the "hairy stars" that were messengers from the gods).

It is worth asking whether Eudoxus actually believed in the physical reality of his system of spheres. Eudoxus was a mathematician as well as an astronomer. What little evidence that now exists suggests that he considered his theory to be a mathematical system, not necessarily physically true. For example, it is an observational fact that the planets vary in brightness over time. The Eudoxian spheres, however, make no allowance for this. Planets fixed on a perfect sphere would not so vary. Another problem with the physical reality of such spheres is mechanical. The outermost spheres would transfer some of their movement to the inner ones. They would need counterbalancing spheres, moving in the opposite direction, to cancel out those motions. Eudoxus made no mention of such additional spheres (which would have greatly multiplied the total number of spheres needed). Finally, Eudoxus seems to have never addressed the actual composition of the spheres. He seems to have regarded them, and the entire system, as a mathematical model and nothing more. Nevertheless, the Eudoxian spheres stand as the first serious attempt at a scientific theory of the heavens and the movement of celestial bodies.

Eudoxus's system was later modified a bit by Callipus of Cyzius (c. 370–c. 300 B.C.). By adding two additional spheres for the sun and the moon, and one additional sphere for Mercury, Venus, and Mars, Callipus increased the mathematical accuracy of the system. This modified version of Eudoxus's spherical cosmology included a total of thirty-four spheres.

The Great Teacher

Aristotle (384–322 B.C.) was born while Eudoxus was still alive. He was certainly aware of Eudoxus's work and adopted a modified version of his system of spheres. There was one major philosophical distinction between the system of Eudoxus and that of Aristotle. While Eudoxus probably considered his theory as only a mathematical model, Aristotle felt that it was a *physical reality.* The crystal spheres, he said, were made of a perfect transparent material, and the motions of the spheres were produced by the Prime Mover, one who acts from beyond the universe. Aristotle recognized that more spheres were necessary for the scheme to be physically real, and he added them to his system. The additional spheres were counterrotating. They canceled out the effects on the inner spheres of the motions of the outer spheres. Aristotle's scheme had a total of fifty-six spheres.

Like his teacher Plato and other Greeks, Aristotle associated certain gods with specific planets. Unlike the Babylonians, he and other Greeks did not literally *identify* a planet as a god. Venus, for example, was a wandering star that also had a significant association with Aphrodite. But it wasn't the goddess herself.

ARISTOTLE'S SPHERICAL SYSTEM

Celestial Object	Number of Spheres
Stars	1
Saturn	4
Jupiter	4
Counterspheres	3
Mars	5
Counterspheres	3
Venus	5
Counterspheres	4
Mercury	5
Counterspheres	4
Sun	5
Counterspheres	4
Moon	5
Counterspheres	4
Total	56

More significantly for Western philosophy, religion, and science, Aristotle had some strong opinions about the nature of celestial motions. The cosmos, thought Aristotle, was divided into two realms. The realm below, including the Earth, was one of change. The realm above, which was the heavens, was one of eternal perfection. Motions in the heavenly realm, he said in his book *Physics*, must all be circular. The reason is that a circle is a perfect form, an eternal form. He distinguished between the eternal perfect realm of the heavens and the highly imperfect realm of Earth. It was obvious that heavenly objects must move in circles, since that is the realm of perfection. What's more, there was no force needed to make heavenly objects or the celestial spheres move. They simply *moved*, since they were perfect. Things were different in the terrestrial realm of earth, air, fire, and water. Here on Earth, objects move toward their "natural place." Objects associated with fire and air move upward toward their natural place, the sky. Objects made of or associated with earth and water move downward, toward the center of the Earth.

In that regard, Aristotle (along with most other people of his time) did not believe that the Earth was flat. He knew it was a sphere. No one who was part of an active seafaring culture could help being aware of what happens to ships as they appear on the horizon and sail closer to port. First the tops of the masts appear, then the rest of the masts, and then the ship. Clearly, the Earth's surface is curved. In his work *On the Heavens* Aristotle also noted that the shadow of the Earth upon the moon during a lunar eclipse is circular. That at least implied that the Earth was a sphere. Even more to the point, he observed that the southern constellations gradually rose higher above the horizon the farther south one traveled.

Although the natural movement of these elements was toward the Earth's center, motions that involved force could also take place. To keep a wheel moving, for example, someone must continually push or pull it. Once that force stops, the wheel stops moving. Aristotle thus concluded that these motions needed a force to keep them going.

While Aristotle believed the Earth was round, he also was convinced that it was stationary and lay at the physical center of the cosmos. If the Earth were moving, he reasoned, then objects thrown into the air would not come back to the ground, as their natural movement would require. That they did was proof that the Earth was stationary. He also reasoned that the lack of any stellar **parallax** proved that the Earth was at the center of the cosmos. Parallax is the apparent motion of a celestial object when it is observed from two widely separated points. Suppose that the stars are attached to a crystal sphere, but that the Earth circles around the sun, not the sun around the Earth. Now find two stars close together in the night sky and at midnight note their distance from each other. Three months later, just after sunset, make the same observation. The Earth has moved away from them in its path around the sun, and the stars will appear closer together. Now wait six months and make a third observation just before sunrise. The stars will still appear to be closer together than they were nine months earlier. We wait three

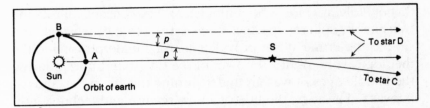

An example of heliocentric stellar parallax: when Earth is at A, the nearby star S lines up with the more distant star D. Three months later Earth is at point B in its orbit around the sun. Now star S aligns with star C. The angle between stars C and D, or p, is also the angle formed at nearby star C by the motion of the Earth. Angle p is the heliocentric parallax of star S, and astronomers can use it to determine the actual distance from the Earth to the star.

Illustration from Robert T. Dixon, *Dynamic Astronomy*, fifth edition, 1989. Reprinted by permission of Prentice-Hall, Inc., Englewood Cliffs, New Jersey.

months and make one more observation at midnight. The two stars appear now farther apart than they were three and nine months earlier, and the same distance apart as one year earlier.

This shift in angular position is called heliocentric stellar parallax. If the Earth were in orbit around the sun, and the stars were relatively close to the Earth, this is what we (and the ancient Greeks) would see. But they didn't see this at all. Aristotle's model of the universe explained why there was no heliocentric stellar parallax—the Earth is at the center of the universe, and it is not moving. This model was accepted as correct, because it explained in a natural way what people could see (and not see) with their own eyes. The reason it was not in fact correct was that Aristotle and everyone else at that time made a false assumption and an inaccurate observation. The false assumption was that the stars are relatively close to the Earth. They are not. The inaccurate observation was that heliocentric stellar parallax does not take place. It does—but it is too small to be seen with the naked eye.

Aristotle's writings were lost to the European world for many centuries. However, they were eventually found and translated from the Greek by Arabic scientists of the early Islamic world. From there they made their way back into European philosophy and theology, largely through the writings of Saint Thomas Aquinas. The Aristotelian worldview strongly supported that of Ptolemy (more about him later). The concepts of rotating transparent spheres, perfect circles, the eternal heavens, and God as the Prime Mover provided a strong underpinning to the theological and political power of the Christian Church during the Middle Ages. It was an underpinning that would eventually be washed away by the tide of a new science and technology.

The Alexandrian Trio

Aristotle had many pupils, but his most famous was the young Macedonian king Alexander the Great (356–323 B.C.). Alexander's armies swept from Greece southwest to Egypt and as far east as the fabled Indus River, spreading Greek thought and philosophy across the known world. Alexander himself died of a fever at the age of thirty-three, but his legacy lived on in the kingdoms overseen by his generals. In 307 B.C. one of them, Ptolemy Soter, founded a great library in the Egyptian city built by Alexander in 332 B.C. and named in his own honor. The library at Alexandria became the center of Greek learning. And it was there, in the city named for the charismatic Macedonian king, that three of the greatest Greek astronomer/philosophers to have an impact on the cosmology of the Western world lived and carried out their work.

Aristarchus of Samos (c. 310–c. 230 B.C.) is best known for his radical heliocentric model of the cosmos. Aristarchus believed that the sun, not the Earth, was the center of the universe. The Earth, the moon, the planets, and all the stars traveled in circular paths around the central sun. He also taught that the spherical Earth rotated on its axis once a day, thus explaining why the sun appears to rise and set and why the stars seem to move across the night sky as if attached to a distant transparent sphere. The Earth moved around the sun in a circular path once a year, and that explained the annual movement of the sun through the zodiac.

Aristarchus apparently wrote many "books" (actually, rolls of papyrus; the book as we know it today had not yet been invented) containing his philosophical and scientific speculations. His writings were stored in the library at Alexandria. They were known to and read by many other thinkers of his time and later ages, and were commented upon by several.

A second important Alexandrian philosopher/astronomer of the third century B.C. was Eratosthenes of Alexandria (c. 276– c. 195 B.C.). Eratosthenes is best known as the first person to make an accurate estimate of the size of the Earth. He estimated based on two assumptions and one observation. First, he believed that the sun was far enough away from the Earth that its rays could be treated as if they were parallel. Eratosthenes also knew that the Egyptian cities of Alexandria and Syene were 5,000 stades apart; a **stade** is thought to be about 148.8 meters. (Syene is today known as the city of Aswân, and is the site of the famous Aswân High Dam on the Nile River.) He also assumed that Syene lay due south of Alexandria and thus on the same meridian, which is any line that runs due north-south, from the North through the South Pole.

Eratosthenes had been told by a friend that in Syene at high noon on one day each year a tall pole perpendicular to the ground casts no shadow. The following year, on the same day, Eratosthenes stuck a pole in the ground in Alexandria. At high noon he noticed that the pole cast a shadow that was one-fiftieth its length. He therefore (correctly) concluded that the distance from Alexandria to Syene must be one-fiftieth the circumference of the Earth. Therefore, the Earth must have a

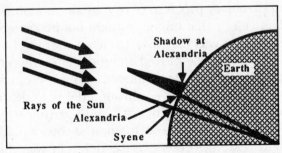

In a brilliant use of mathematics and scientific observation, Eratosthenes of Alexandria deduced the circumference of the Earth more than a hundred years before the birth of Christ.

Illustration from Michael J. Crowe, *Theories of the World from Antiquity to the Copernican Revolution*, Dover Publications, Inc., 1990. Reprinted by permission of the publisher.

circumference of 50 × 5,000 stades, or 250,000 stades. That works out to about 37,200 kilometers or 23,100 miles—amazingly close to the modern measurement of 40,074 kilometers. The only reason Eratosthenes missed by just 7 percent was that one of his two assumptions and his one measurement were slightly off. Alexandria and Syene are a bit more than 5,000 stades from each other, and they do not lie exactly on the same meridian.

A third important Greek cosmological thinker of that time was Hipparchus of Nicaea (c. 160–c. 127 B.C.). Like Pythagoras some four hundred years earlier, Hipparchus turned to the ancient Babylonians for inspiration, collating and organizing their astronomical observations with the aim of developing a better cosmological model of the universe. He added those observations to the basic cosmology of Aristotle and realized that many nagging inconsistencies in that picture needed to be resolved. They included such things as the retrograde motion of the planets and the nonuniform movement of the sun and the planets through the signs of the zodiac. Observations of the planets clearly showed that they did not move on circular paths at uniform speeds. Yet Aristotle's system required that they do so. To reconcile the planets' actual behavior with philosophical imperatives, Hipparchus invented the geometric devices called **eccentrics**, **epicycles**, and **deferents**.

Hipparachus used epicycles and deferents to explain away the nagging planetary problem of retrograde motion. The deferent was a large circle that might or might not be centered on the Earth. The epicycle was usually a circle smaller than a deferent. The center of an epicycle moved along the circumference of the deferent, and the planet itself moved along the epicycle's circumference. As a planet moved through the complex combination of epicycles and deferents, it would at times be moving opposite to the motion of the deferent and thus would be moving backward. The movement of the deferent was the planet's general movement through the zodiac from west to east.

The eccentric was supposed to explain the other nagging problem, the fact that planets move more quickly through some zodiacal signs than others. According to Hipparchus, a planet like Mars moved in a perfect circle with uniform velocity around the center of that circular path, but the Earth *was not quite at that center*. It was slightly displaced from it. The resulting effect was that the planet *appears* to be moving more quickly through some parts of its path than through others. This, however, was merely an illusion caused by the eccentric.

Interestingly enough, in resorting to the geometric device of the eccentric, Hipparchus *was displacing the Earth from the center of the cosmos*. Not by much, of course, certainly not to the degree that Aristarchus did with his heliocentric cosmology. But there it was. Some 250 years later, a Greco-Egyptian astronomer would adapt Hipparchus's system and turn it into the ruling astronomical paradigm. In doing so, he would include the eccentric, with its geometry of an Earth displaced from the center. It was another small step toward the paradigm shift that would not only completely remove Earth as the center of the universe, but would eventually abolish the idea of a cosmic center entirely.

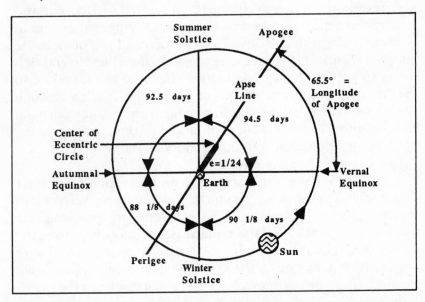

In order to more accurately explain the apparent movement of the sun around the Earth, Hipparchus invented the eccentric and slightly displaced the Earth from the center of the universe. Ptolemy later used Hipparchus's device in his own astronomical system.

Illustration from Michael J. Crowe, *Theories of the World from Antiquity to the Copernican Revolution*, Dover Publications, Inc., 1990. Reprinted by permission of the publisher.

The Ptolemaic Model of the Cosmos

Little is known about the life of Claudius Ptolemaeus, or Ptolemy. We do not even know the year he was born or died. We do know that he lived during the second century A.D.; by backtracking his published astronomical observations, we know that he was working sometime around A.D. 125. Ptolemy was living and working in Alexandria and had access to the centuries of astronomical records and writings that were stored there. His great achievement was a thirteen-volume treatise entitled *Almagest*. In it was a catalog of more than a thousand stars, considerable mathematical information—and his detailed creation of a geometric, geocentric cosmology. The aptly named **Ptolemaic system** became the reigning astronomical paradigm of the Western world for more than fourteen hundred years.

Ptolemy's system was based on Aristotle's model: the Earth lay at the center of the universe; it was a sphere; it did not move; and the celestial spheres carrying the sun and the planets moved in a uniform manner. Then he began modifying that system in order to make it fit the obvious movements of the planets. He adopted Hipparchus's geometric devices, the epicycle and deferent, to explain retrograde motion. He used the eccentric to explain the nonuniform speeds of the planets moving through the zodiac. Then Ptolemy did something truly revolutionary—at least in terms of the Aristotelian model of the cosmos. To explain some additional retrograde motions, he invented a geometric device called the **equant**.

The equant is a circle similar to an eccentric. A planetary object would travel along the equant's circumference. The Earth was offset from the center of an eccentric. In somewhat the same way, a point called the **equant point** was offset from the center of an equant in the direction exactly opposite to that of the Earth, and for exactly the same distance. Now suppose, said Ptolemy, that a line is drawn from this equant point to the planet that is moving along the equant's circumference. As the planet moves along the equant, that imaginary line will move through equal angles in equal times. Because the planet is variously closer and then farther away from the equant point at different points along the circumference, it must *not* be moving at a uniform speed along the equant.

With the invention of the equant to explain retrograde motion, Ptolemy violated one of the primary tenets of Aristotle's cosmology. And he did so for a very good reason: *It was the only way he could manage to make his system conform reasonably well to observations of the planets' actual motions in the sky.* This was a very important step for Ptolemy to take.

His action, in fact, lies at the very heart of science as we know it today. For the real driving force of science is not so much to confirm an explanation or theory, but to find ways to tear it apart. The best way to do that is to test the theory against actual observations of the real world or (if possible) with experiments. If the theory or explanation or principle does not accord with reality, then it must be either modified or thrown out. Aristotle's inviolable rule was that all celestial motions must be uniform. It did not match reality. So Ptolemy modified it. Eventually this Aristotelian tenet would be tossed into the trash can of history. But for then, Ptolemy's move was a good one, in the right direction. The direction of truth.

When finished, Ptolemy's model of the cosmos was fairly simple. However, it did treat some of the planets differently than others. Mercury and Venus somehow had to be forced to stay near the sun, as they do in real life, so Ptolemy fiddled with their deferents to make them do so. He also had to modify the epicycles of Mars, Jupiter, and Saturn to explain the fact that they go retrograde only when they are at *opposition*—that is, when they are at a point in the sky that is 180 degrees from that of the sun.

Another interesting aspect of Ptolemy's cosmology was that he considered the heavenly spheres to be real, physical things. They were made, he thought, from a substance called **quintessence**. The word literally means "fifth essence," distinguishing this substance from the four classical Aristotelian essences of earth, air, fire, and water—the four fundamental elements from which, Aristotle theorized, the cosmos is made.

Ptolemy also took a stab at estimating the size of the universe. Hipparchus and others had set the Earth-moon distance at about fifty Earth radii, and Ptolemy worked from that figure. He assumed that the celestial spheres were all nested tightly together, and he made allowances for the room needed for the planets to move along their various epicycles, deferents, and equants. He ended up placing the sphere of the fixed stars about twenty thousand Earth radii distant. That's not much more than the distance we know today from the sun to the planet Venus. Ptolemy's universe was very small indeed. But it was the universe that the Western world embraced as its own for more than fourteen centuries.

Finally, it's important for us to realize why the Ptolemaic model was accepted as the truth. It wasn't because people in those days were stupid or unobservant. They were as intelligent as we are and as aware of the world they lived in as we are today—perhaps more so. The Ptolemaic system was accepted *because it worked.* And it worked well. It allowed astronomers to predict the positions of the planets with the accuracy they wanted. It agreed with and more or less supported the prevailing Greek-based system of philosophy and esthetics. As Christianity rose from an obscure Jewish sect to the dominant religious system of the West, its leaders and teachers were able to use the Ptolemaic model to confirm and support their own vision of the cosmos, and of humanity's relation to God.

In other words, the Ptolemaic model of the cosmos came to fulfill the major tasks of a functioning mythology. It would not be overthrown until the Reformation, when the Christian Church itself was irreparably torn apart. The revolution against the Ptolemaic system and the Aristotelian worldview would be led by a Polish Catholic cleric, a German who had once attended a Lutheran seminary, and an Italian who numbered among his friends the Pope.

4
THE GANG OF FOUR

It was the year 1543. In fire and blood the Christian Church had broken apart. King Henry VIII of England was getting ready to marry Catherine Parr, his sixth wife. In Spain the Inquisition began burning Protestants at the stake, and Ignatius of Loyola was starting the Society of Jesus. A new pope in Rome took the name Paul III. Titian was painting *Ecce Homo* ("Behold the Man"). And the Portuguese had landed in Japan, bringing with them the "gift" of firearms. Nearly five hundred years would pass before a priest and poet named Karol Wojtyla would gain fame as the first pope from Poland. But another Polish cleric was about to gain his own measure of immortality by shaking the Western mythological foundation of the heavens into a new shape. His name was Nicolaus Copernicus (1473–1543).

Copernicus and the Sun-Centered Cosmos

Copernicus studied law and medicine in Italy, where he came across the works of Aristotle, Pythagoras, and Plato. The young Pole was particularly fascinated with the philosophical model called Neoplatonism, which was based on some of Plato's teaching. Among other things, Neoplatonism stressed the uniqueness of the sun, which was considered to be the source of all knowledge and the seat of the Godhead itself.

Copernicus began to consider a model of the cosmos different from the geocentric Ptolemaic one, a heliocentric model— that is, one with the sun at its center. Not that he was bothered by any obvious shortcomings of the Ptolemaic model. At that time, there seemed to be very few. The Ptolemaic model worked very well as a predictive tool—which is what we expect of a good scientific theory. In fact, the cosmological model that Copernicus was to develop was at first no better at predicting planetary positions than that of Ptolemy. What inspired Copernicus to think about the cosmos in a new way was the same kind of thing that has inspired many other scientists: it was his intuitive sense of esthetics, harmony, order, beauty. It simply seemed to Copernicus (influenced as he was by Neoplatonic ideas) that a sun-centered cosmos was more *elegant* than one with the Earth at the center.

Copernicus first presented his ideas in a short unpublished manuscript entitled "Commentariolus," written in 1512 when he was about forty years old. He had by this time taken a position as canon (a clergy position similar to that of an associate minister) at the Catholic Cathedral in Frauenberg, East Prussia. Some seventeen years later he wrote another summary of his thesis and circulated it among some friends. His heliocentric hypothesis was essentially the same as that of Aristarchus some eighteen hundred years earlier. Copernicus asserted that the sun was at the center of the solar system and that the planets— including the Earth—moved around it. The Earth took one year to make its way around the sun, and it also rotated on its own axis once a day. Copernicus believed that the orbits of the planets were perfect circles and that Ptolemy's equant was in violation of that cosmic ideal. The problem with the Ptolemaic model, he wrote, was that it was "not sufficiently pleasing to the mind."

Copernicus continued working on his theory, slowly writing a more complete treatise, which he titled *De Revolutionibus* ("On the Revolutions"). Sometime around 1542 or 1543, Copernicus's earlier, privately distributed work came to the attention of the German astronomer Georg Rheticus. He met with Copernicus, who showed him a completed manuscript copy of *De Revolutionibus.* At Rheticus's urgings, he agreed to allow the manuscript to be printed. (The first printed Bible had been produced about twenty years before Copernicus's birth.) Not long before the book was published in April 1543, Copernicus

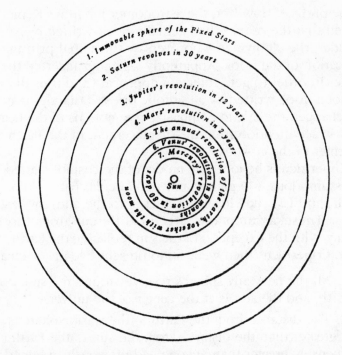

In the Copernican astronomical system the outermost sphere was that of the immovable stars, followed by the spheres of Saturn, Jupiter, Mars, the Earth and its moon, Venus, and Mercury. At the center of the cosmos was the sun. "For in this most beautiful temple," wrote Copernicus, "who would place this lamp in another or better position than that from which it can light up the whole thing at the same time?"

Illustration from Michael J. Crowe, *Theories of the World from Antiquity to the Copernican Revolution*, Dover Publications, Inc., 1990. Reprinted by permission of the publisher.

suffered a serious stroke. A copy was delivered to him, but it is likely he died without having read it.

There are two curious asides to the story of Copernicus's revolutionary *De Revolutionibus*. The first is that the title was changed to *De Revolutionibus Orbium Coelestium* ("On the Revolutions of the Heavenly Spheres"). Second, an unsigned preface was added, which essentially said, "This is a hypothesis that some might find interesting, but I don't want people to think I believe the cosmos is really like this." For many years people thought that Copernicus had changed the title and written the preface. However, the astronomer Johannes Kepler discovered that the preface had actually been written by Andrew Osiander, the clergyman who oversaw the actual printing and publication of the book. Apparently Osiander inserted the disclaimer in order to get Protestant approval (and thus distribution) of a book written by a Catholic priest. Osiander may also have changed the title to emphasize the motions of the heavens and distract attention from the displacement of the Earth from the center of the cosmos.

Copernicus's heliocentric model of the cosmos started with two assumptions: (1) planets move in circular paths at constant speeds, and (2) the closer a planet is to the sun, the faster it moves. These assumptions are no different from those of Ptolemy, with the exception of the sun replacing the Earth at the center. Copernicus then went on to present five major ideas:

1. All the heavenly spheres revolve around the sun, not the Earth, and the sun is at the center of the universe.

2. The distance from the Earth to the starry sphere is much greater than the distance from the sun to the Earth and is immensely greater than anyone had previously assumed.

3. The heavens appear to move around the Earth in their daily pattern due to the rotation of the spherical Earth upon its axis.

4. The apparent motion of the sun through the zodiac is caused by the Earth's annual orbit around the sun.

5. The retrograde motions of the planets in the sky are illusions and are actually caused by the combination of their orbital motions around the sun with that of the Earth.

This last point could use a little elaboration. Suppose the Earth is a car driving down a long curving road on perfectly balanced shocks. It's moving smoothly and steadily. You are a passenger in the car. The planet Jupiter is another car traveling along a second curving road far to your right. Off in the distance, far behind the Jupiter car is a range of mountains silhouetted against the afternoon sky. Your car is behind the Jupiter car, but you are moving much faster. You start catching up with the Jupiter car, and eventually pass it. Now, while you are be-hind it and catching up with it, the Jupiter car will appear to be *moving backward in reference to the distant mountains.* You catch up with it and pass it by, and with reference to the moun-tains, the Jupiter car appears again to be moving forward. This is the illusion of retrograde motion. Copernicus didn't know anything about automobiles, but he knew about retrograde motion.

Copernicus was able to calculate the sidereal periods of the planets—that is, the orbital periods of the planets with respect to the stars, as seen from the sun. From this he found that the order of the planets fit their sidereal periods. Mercury, with the shortest period, was the closest to the sun, followed by Venus, Earth, Mars, Jupiter, and Saturn. He also calculated the relative distances from the sun to each planet, using the distance from the sun to the Earth as the unit of distance (a measure that is now called an astronomical unit, or AU). His calculated dis-tances were quite close to the actual ones.

Copernicus's model was far from perfect. Though he threw out Ptolemy's equant and epicycles, he still had to use his own additional circles and versions of epicycles to explain the nag-ging variations in the planets' orbits. Copernicus's model was in some ways actually more complicated than that of Ptolemy.

Also, Copernicus did not introduce the idea of a force acting between the sun and the planets. Like Aristotle and others, Copernicus believed that the motions of the planets arose from the motions of celestial spheres, motion without any force.

The real impact of the Copernican heliocentric model was not in its greater predictive power (it had none), nor in its greater geometric simplicity (it didn't have that either). Rather, it lay in its appeal to a harmony of cosmic design and to a "fixed symmetry of its parts," as Copernicus wrote. And its appeal lay in its firm grounding in observation. Some eighteen hundred years earlier Aristarchus had come up with much the same cosmological vision. But for Aristarchus a heliocentric cosmos was an interesting *idea.* It was not meant to be taken as some-thing real. By contrast, and in contrast to the unauthorized preface at the beginning of *De Revolutionibus Orbium Coelestium*, Copernicus's heliocentric model was always a model for the reality.

Sixty-six years after the publication of *De Revolutionibus Orbium Coelestium*, a German Lutheran astronomer would improve on the vision of the Polish Catholic priest.

Kepler's Three Laws

It was the year 1609. In England the popular playwright William Shakespeare had begun writing *Cymbelline*. Construction of the Blue Mosque began in Constantinople. The Dutch East India Company shipped its first load of Chinese tea to Europe. Henry Hudson was busy exploring Delaware Bay. Galileo Galilei was playing with his newly constructed *occiale*. And Johannes Kepler was wondering what kind of force from the sun could cause planetary motions. In particular, he was wrestling with the orbit of Mars.

Born two days after Christmas in the town of Weil der Stadt in southwestern Germany, Kepler (1571–1630) was a brilliant student at the University of Tübingen. He went on to become a teacher at a high school in Graz, Austria. It was there he came to believe in a geometric model for the solar system. He had learned of the Copernican theory of the solar system while at Tübingen, and in Graz began working on ways to incorporate it into his vision of a geometric-ruled cosmos. In particular, Kepler wanted to answer three questions: (1) Why are there six planets? (2) Why are their orbits positioned as they are? (3) Why do planets farther from the sun move more slowly than those closer to the sun?

In answer to the first two questions, Kepler hit upon a theory based on what he felt was a relationship between the spacing between planets and the five classical or so-called Platonic geometric solids. As we saw in the preceding chapter, these are the tetrahedron, cube, octahedron, dodecahedron, and icosahedron. The orbits of Mercury and Venus, Kepler theorized, were separated by an octahedron. Venus and Earth were separated by an icosahedron; Earth and Mars, by a dodecahedron; Mars and Jupiter, a tetrahedron; Jupiter and Saturn, a cube.

Kepler was quite serious about his geometric theory of the

solar system. He was convinced that his planet-spacing tetrahedrons and dodecahedrons really existed, and that they explained both the number and positioning of the planets. To support his erroneous theory with numbers, Kepler thoroughly studied the motions of the planets around the sun. This study in turn convinced him that the sun exerted some kind of force on the planets, which drove them in their orbits, with the closer planets moving more quickly than those farther away.

In 1600 Kepler (a devout Lutheran) fled a Counter-Reformation crackdown in Graz and traveled to Prague, where he joined the staff of his personal hero, Tycho Brahe (1546–1601), the Dane who was the last great pre-Copernican astronomer. Though their personalities clashed so much that they could not work together, Brahe and Kepler were each impressed with the other's abilities. Just before he died, Brahe urged Kepler to carry on his work using his (Brahe's) unpublished astronomical observations. Kepler began this task by trying to model the movement of Mars using those observations and a heliocentric scheme. It didn't work very well. In the process, however, Kepler realized that the closer Mars was to the sun, the faster it moved, and vice versa. He was soon to codify this insight as one of his famous three laws of planetary motion.

By 1609 Kepler was still trying to figure out the nature of the solar force driving the planets. At this point he was inspired by a book written by the English philosopher William Gilbert, *De Magnete*, and decided that the sun was a fountain of magnetic forces that drove the planetary motions. He then applied this idea to his ongoing attempt to model the motions of Mars. It was then that Kepler realized that the only way his model would work was if Mars followed a path around the sun that was not circular, but elliptical. This was another one of Kepler's laws of planetary motion. The third and final of his three laws was formulated nine years later.

These three laws of planetary motion would later be a key ingredient in Isaac Newton's discovery of the law of universal gravitation. They are:

1. *The Law of Ellipses*—The shape of the orbit of each planet is an ellipse, with the sun at one focus and the second focus empty.

2. *The Law of Equal Areas*—An imaginary line drawn from the sun to a planet traveling along its elliptical orbit will sweep across equal areas in equal amounts of time. Thus, the orbital velocities of a planet are not uniform, but they do vary in a regular way: the closer a planet is to the sun, the faster it moves, and the farther from the sun it is, the more slowly it moves.

3. *The Harmonic Law*—The *square* of a planet's orbital period is directly proportional to the *cube* of its average distance from thé sun. This suggests that planets with larger orbits, which are thus more distant from the sun, move more slowly in those orbits than planets that are closer to the sun.

It was Kepler's third law that played a pivotal role in Newton's discovery of universal gravitation, for it implies that the sun-planet force (whatever it is) decreases with distance.

In his treatise *Astronomia Nova* ("The New Astronomy"), Kepler presented convincing evidence for his theory of elliptical planetary orbits and the existence of a force between the sun and the planets. In doing so he broke finally and forever with the ancient Aristotelian and Ptolemaic cosmologies of perfect cosmic spheres and spherical orbits. An old mythology was dying, and a new one was about to be born—one that would lead inexorably to a vision of a sea of stars, a cosmos of galaxies, and holes at the center where time and space come to a stop.

Galileo Reveals a New Cosmos

Galileo Galilei (1564–1642) can rightly be considered the godfather of modern physics, astronomy, and optics. His breakthroughs in those areas of scientific inquiry put the Copernican cosmology on firm observational footing, paved the way for Newton's cataclysmic rearrangement of the Western worldview, and turned the telescope from an object of mild amusement into an instrument of revolution. The publication of his *Siderius Nuncius* ("The Starry Messenger"), the first popular astronomy book, also made Galileo the first mass media "pop scientist," paving the way for people like Harlow Shapley, Isaac Asimov, and Carl Sagan.

The year of Galileo's birth was vintage. It also saw the birth of William Shakespeare, Christopher Marlowe, and Pieter Breugel the Younger, and the death of Michelangelo and John Calvin. It was also the year that the Spanish conquerers of the Philippine Islands began building the city of Manila, that the horse-drawn coach was introduced into England from Holland; and that the notorious *Index Librorum Prohibitorum* ("The Index of Forbidden Books") was published with the approval of Pope Pius IV.

On February 15 of that year, Galileo Galilei was born in Pisa to parents originally from Florence. Galileo's father was a well-known musician, and the family was moderately well-off. When Galileo was ten, they moved back to Florence. A year later he began a period of study at the monastery of Santa Maria near the town of Vallombrosa, and at one point became a novice of the religious order there. However, at age seventeen he left the monastery to study medicine at the University of Pisa. Two years later he became fascinated with mathematics and geometry, and abandoned his medical studies. He left Pisa for home in Florence, where he spent several years teaching, lecturing, and writing, but again returned to Pisa in 1589 for a three-year stint as a professor of mathematics at the university.

It was during this time that Galileo began studying the Copernican heliocentric system. He also wrote of his intention to use mathematics to study natural phenomena. His contemporary, Kepler, was using mathematics as a way to uncover the harmonies of the cosmos. Galileo, however, was thinking in more concrete terms: mathematics would be a tool in his hands to dissect and understand concrete problems.

In 1592 his appointment ended, and Galileo was unemployed. However, a friend and patron got him a job teaching astronomy and geometry in Padua, where Galileo would live and work for the next eighteen years of his life. They would be his most scientifically productive ones. At the university he found a harmonious atmosphere, and Padua was a quiet and friendly city. Galileo did liven things up at times by becoming embroiled in several professional quarrels. Galileo had a sharp tongue and a savage wit, and the "scientific" style of discourse in the seventeenth century was more like barroom brawling than intellectual jousting. (In truth, that is still often the case. Anthropologists still regularly incorporate personal insults into their controversies, for example, and the "debate" over who really discovered the AIDS virus has not been exactly a model of decorum.)

Family problems caused Galileo considerable emotional and economic stress, but he managed to get through them. And in 1599 he set up housekeeping with the love of his life, a Venetian woman named Marina Gamba. Although they never married, Galileo and Marina had three children.

By 1597 Galileo had become convinced of the correctness of the Copernican vision of the cosmos. He wrote to Johannes Kepler (then still in Graz) to tell him of his scientific conversion. But nothing much would come of this conviction until the summer of 1609. A messenger on his way to Venice stopped in Padua to tell Galileo that a man in Holland named Hans Lippershey had invented a curious tool made of an open tube with glass lenses at each end. This spyglass made distant objects appear closer than they really were. Galileo immediately set about building his own *occiale* (the name "telescope" was later given it by Federico Casi). His first telescope was a crude one; it magnified objects only thirty times. But that was enough for starters.

Galileo was nothing if not a practical man. He immediately recognized the military potential of this new instrument. He wrote about it in a letter to the Doge and Senate of Venice, rulers of the Italian region that included Padua. The politicians decided there was no way to keep this new Dutch invention a secret. The device was already becoming something of a novelty toy throughout Europe—there is nothing new about adolescent boys (and men) using telescopes to peep through the windows of the woman down the road. The Doge and Senate did express their thanks to Galileo for tipping them off to the telescope's military uses: they raised his salary to one thousand florins per year, an unprecedented amount. Thus, among other accomplishments, Galileo became the first "modern scientist" to make a living as a "military consultant."

However, it is for his scientific discoveries using his *occiale* that we most honor Galileo. When he turned his telescope to the night skies, it provided several amazing discoveries. Not surprisingly, the first celestial object he looked at was the moon. Galileo discovered that its surface was not the smooth, perfect sphere demanded by Aristotle but was wrinkled and pitted, covered with what looked like ocean basins and mountain ranges. He measured the height of one lunar mountain, based on the length of its shadow. The mountain was six kilometers high.

Then he looked at the Milky Way with his telescope—and was utterly astonished. For that band of light that stretches across the night sky, which humans had variously identified as milk, as a river, as a tent seam, was none of these. *It was stars; millions upon millions of stars.* This was another blow to the Aristotelian cosmology, which had held that the cosmos contained only a fixed and unchanging number of stars.

His third astronomical discovery was perhaps even more important than the revelation of the Milky Way's nature. On the evening of January 7, 1610, Galileo looked at Jupiter with his magnifying tube. He discovered several new stars near the planet. Actually, they were four moons of Jupiter. *And they moved.* Galileo tracked their movements through March 2 and drew them on paper. The starry points were clearly moving around Jupiter, and Galileo determined their orbital periods. This was the greatest blow against the prevailing Aristotelian/Ptolemaic worldview. For Jupiter and its satellites, he realized, were a miniature version of the solar system. Here, for every eye to see, was evidence of *a second center to the known cosmos.* That violated the very core of the Aristotelian worldview. Clearly, the Earth was not the center of the universe.

On March 12, 1610, Galileo published a small book entitled *Siderius Nuncius* ("The Starry Messenger"), containing the results of the first astronomical observations ever to be made with a telescope. The public was utterly fascinated. *Siderius Nuncius* became the world's first bestselling science book. It brought Galileo instant fame—and more money, since his private workshop cum laboratory was now swamped with orders for telescopes.

It also brought him considerable criticism. His detractors asserted that his observations were merely atmospheric disturbances or were caused by flaws in the telescope's lenses. True to form, the attacks were often personal and biting. And true to form, Galileo struck back with his own vicious wit.

Despite the objections, Galileo was convinced that his observations were valid and that they proved the truth of the Copernican model of the universe. By 1611 Johannes Kepler was backing up Galileo's observations with his own and with a theory of optics that supported the validity of telescopic observations. Unlike Galileo, Kepler was respected in scientific circles, and his words carried considerable weight.

Meanwhile, Galileo himself continued his astronomical work. His next major breakthrough was the discovery of dark blotches on the surface of the sun. Today we know them as sunspots. It was another hammer blow to the Aristotelian cosmology. The sun was supposed to be a perfect sphere, and it was now shown to be blemished. In 1613 Galileo published these results in a pamphlet entitled "The Letters on Sunspots." For the first time he publicly declared his support for the Copernican model of the universe.

Galileo and Motion

Galileo was not content to stop with his observational restructuring of the universe. He was convinced that the Copernican system needed some kind of physical explanation for the motion of objects. Aristotle's system had one, but the Copernican system did not, in Galileo's view. He was, of course, ignoring Kepler's construction of a mathematical description of celestial motions. Galileo believed that what was needed was a description and explanation of the motions of terrestrial objects. This in turn might be transferred to celestial bodies. Besides, it was easier to do experiments with lead balls than with Mars. So Galileo proceeded to devise new explanations of the motions of falling bodies. In doing so he revised the definition of acceleration, identified the existence of an attractive force between bodies (which we today know as gravity), anticipated Isaac Newton's three laws of motion, and paved the way for Albert Einstein's theory of special relativity.

SPEED, VELOCITY, AND ACCELERATION

We all have experienced speed, velocity, and acceleration, but few of us really have an intellectual understanding of what they are and how they differ.

First of all, **motion** is the change in position of one object with respect to another. I am one object; my apartment is another. I get in my car and drive from my apartment to the local supermarket, which is about a mile away. It takes me about three minutes to get there. I have been in motion.

Also, speed and velocity are not the same thing. **Speed** is the average rate of travel for some object. It is a measurement of an object's movement through some distance in some amount of time. The speed limit on this freeway is 55 miles per hour. The speed of light is 299,792 kilometers per second. **Velocity**, on the other hand, is speed in *some direction*. We drive from New York to Boston at a velocity of 50 miles per hour. We are moving at a particular speed (50 miles per hour) in a particular direction (toward Boston). If we return from Boston to New York at a speed of 50 miles per hour, our speed is the same, but our velocity is different, because our direction has changed.

Velocity and speed are both measured in the same way— the velocity of our trip from New York to Boston is also expressed as 50 miles per hour. In general practice (and—I admit it—in this book), the two terms are often used interchangeably. However, the fact remains that they are different. To reiterate, unlike speed, velocity also involves a *direction* of travel.

Acceleration is a change in *velocity*, not speed. A rocket begins its trip into space at rest, with a speed (and velocity) of zero. (Well, not really; the Earth is rotating on its axis, so the rocket is moving. But let's ignore that for the moment.) Its engines ignite, and it climbs off the launch pad, moving faster and faster in a particular direction. This is acceleration—a change in velocity.

Suppose you are in a race car circling the Indianapolis Speedway at a constant speed of 125 miles per hour. Your speed is the same, but your *velocity is changing*. It is changing because your direction of movement is changing. That means you are accelerating; even though your *speed* is constant, your velocity is not, and acceleration is a measurement of a change in velocity, not speed.

Acceleration is measured as distance per unit of time per unit of time, such as meters per second per second. This is mathematically written as either m/sec/sec, or as m/sec^2 (meters per second squared).

Galileo began by tackling the nagging problem of **inertia**. Inertia is the unwillingness of some object to instantly move when a force is applied to it. Try pushing a large boulder to the edge of a cliff. It resists your force for a while, then it starts to move. That resistance is inertia. By the same token, inertia is the tendency of a body that is already in motion to remain in motion, unless acted upon by an outside force.

Inertia was a problem for Aristotelian physics. Aristotle had divided motions into two types, natural and forced. An example of natural motion was the falling of a rock to the Earth, which was caused by the natural tendency of an earthly object to move toward the center of the cosmos (the center of the Earth). No external force was involved, said Aristotle; this was merely the rock's natural tendency. On the other hand, a rock moving horizontally through the air after being thrown was an example of forced motion. Its motion was "unnatural" and caused by a force being continually applied to it.

Galileo turned this Aristotelian idea on its head by carrying out some experiments with balls rolling down inclined planes. (Some historians have suggested that these "experiments" were actually never physically carried out. Rather, they were what Einstein would later call *gedanken* experiments—thought experiments. More often than not, however, Galileo's conclusions turned out to be correct.) He concluded that a ball rolling along a frictionless surface would continue to move indefinitely as long as no external force was applied. This movement, Galileo said, was caused by the object's own inertia. At the same time, an object resting motionless on a surface would stay motionless unless an external force acted upon it. In reaching these conclusions from his experiments (*gedanken* or actual),

Galileo anticipated Newton's three laws of motion. He also struck another blow against the Aristotelian worldview.

Galileo had long been fascinated by the motion of pendulums, and his observations of their movements led him to another conclusion: once it is in motion, a pendulum will continue to rise and fall without any further input of energy, assuming no external force impedes it. This, too, would later fit in with Newton's laws of motion.

Galileo then applied his idea of inertia to falling bodies. Unlike Aristotle, he claimed that a falling body falls to the ground because it is attracted by an external force—gravity. The object, furthermore, is in constant acceleration. Its velocity changes at a constant rate as it falls. Today we know that this acceleration due to gravity at the Earth's surface is 980 centimeters per second per second. It is denoted as g.

Galileo then took another audacious step by claiming that *every* falling body falls at the same acceleration. This directly contradicted the traditional belief that objects with different masses (lead balls, copper balls, wooden balls) fall at different rates. (It is highly unlikely, by the way, that the legendary "Leaning Tower of Pisa experiment" ever happened.)

Finally, some of Galileo's observations and intuitive leaps led him to anticipate Einstein's eventual destruction of the concepts of absolute space and absolute time. Galileo believed that the laws of nature would be the same on a ship sailing the seas as on one docked at port. It didn't matter how fast the ship was moving. Balls of different masses dropped from the top of the mast would hit the deck at the same time and at the same spot, and would have the same acceleration. This was a relativity theory—a crude one, to be sure, a "Galilean relativity," but relativity nonetheless. Galileo eventually concluded that orientation and position in space, position in time, and the velocity of motion are all relative. Isaac Newton would later reject this in favor of absolute space and time, but he would do so for theological, not scientific, reasons.

The story of Galileo's tragic tangle with the Catholic Church is well known, and I need not go into any detail about it. In brief: In 1615 Robert Cardinal Bellarmine (who would later be canonized by the Church) warned Galileo to present the Copernican cosmology only as a hypothesis, not as fact. The following year the Church's notorious Holy Office placed Copernicus's book on its *Index of Forbidden Books* and warned Galileo to stop defending the Copernican worldview. Galileo eventually was prohibited from carrying out any active experimental work and was banished to Tuscany. In 1621 Bellarmine and Pope Paul V died, and a close clerical friend of Galileo's became Pope Urban VIII. Three years later Galileo went to Rome and persuaded his friend Pope Urban to let him write a book on the Copernican system. For the next eight years, Galileo worked on his masterpiece. The *Dialogue on the Two Chief World Systems, Ptolemaic and Copernican,* was published in 1632. It vigorously defended the Copernican system and presented Galileo's theory about the relativity of motion and his concept of inertia. The following year the excrement hit the fan. Galileo was brought to ecclesiastical trial before the Holy Office, accused of heresy. The *Dialogue* was banned, and Galileo was forced to publicly recant his belief in the Copernican system. He died on January 8, 1642.

Some 350 years later, the Catholic Church's leaders have apologized for the treatment of Galileo and have talked about posthumously throwing out his ecclesiastical convictions. That has yet to happen. However, it doesn't matter all that much. Three and a half centuries have revealed the enduring truth of Galileo's work. Time has also shown the error of the Church's insistence on trying to harness the validity of spiritual truths to a particular cosmology. For cosmologies change, and our mythological visions of what lies at the center of the universe change with them.

The Newtonian Revolution

Mark Twain once commented, "Everybody talks about the weather, but nobody does anything about it." Something similar could be said of gravity. People have always known about gravity. It is something we experience every day and sometimes even notice. I drop an apple; it falls to the ground: gravity. You upend a pitcher of mead over your husband's head; he is drenched: gravity. He throws his enemy over the cliff; the man falls to the rocks below and dies: gravity. However, no one—not even Aristotle—ever did anything scientific about it until Isaac Newton.

Our current-day cosmology, with its realization that the universe is *very* big, that it contains cities of stars we now call galaxies, that we live inside a galaxy, that we are nowhere near the center of our Galaxy, that things called black holes can exist, that the center of the Galaxy may be home to a black hole so huge it could swallow a million or more suns and only double in mass—all these aspects of our current worldview were made possible by Isaac Newton. In a very real sense, Newton discovered gravity. And in doing so he changed the way we perceive the cosmos and our place within it.

Newton (1642–1727) was born on Christmas Day in the same year that Galileo died. He was a frail infant whose father died shortly after Newton was born. When he was two years old, his mother remarried and got rid of him, sending him off to be raised by his maternal grandmother. It was the kind of stressful and neglected childhood that often leads to the growth of a neurotic or even psychotic adult. Newton was a classic example. As a teenager he briefly showed interest in a neighbor girl but later dropped it because (he is supposed to have said) such an involvement would only take time away from his career. Newton never formed another attachment with a woman in his life and probably died a virgin. He suffered panic attacks at the very thought of his work being published. He often flew into a violent rage when his ideas were challenged. He had at least two nervous breakdowns. He had several episodes of severe mental disturbance during the latter years of his life. He also exhibited other forms of bizarre behavior at various times. Some may have been due to mercury poisoning from his experiments in alchemy, a subject with which Newton was fascinated.

It may seem peculiar to those of us immersed in the materialistic cosmology of our times that Newton's greatest enthusiasms—aside from mathematics and thinking—were alchemy and the occult. Yet it was not unusual for him, nor was it for his day. The seventeenth century was a period of transition between the cosmologies of the Middle Ages and those of our times. And it was Isaac Newton himself who was instrumental in causing that sea change.

Newton's first career was as manager of his late father's farm in Lincolnshire, and he was an abysmal failure at it. An uncle connected with Trinity College at Cambridge rescued him from an ignominious life as a gentleman farmer and got him into the university. After a fairly undistinguished sojourn in academia, he received his degree in 1665, just as the Black Plague swept through London. It was followed the next year by a great fire, which almost completely destroyed the city. Newton took refuge by moving back to the family farm in Lincolnshire. It was there he made the scientific discoveries that assured his place in history.

First, he began developing the mathematical discipline called calculus. The German scientist Gottfried Wilhelm Leibniz discovered calculus at the same time as Newton. Several years later a bitter controversy raged throughout Europe over which of the two would receive credit as the "official" discoverer of this powerful mathematical system. It is believed that Newton actively encouraged his supporters in their ofttimes virulent attacks on Leibniz.

Calculus was but one of Newton's breakthroughs during the three-year period from 1664 to 1667. He was also interested in optics. In a small room on his farm that was completely dark but for a pinhole through a curtain, Newton experimented with pencil-thin beams of light passing through a triangular glass prism. He discovered that white light is actually made of many colors of light, ranging from red through deep violet. Newton had discovered the **spectrum**.

But that was not all. Newton had learned about Kepler's laws of planetary motion while at Cambridge. Now he took them one step further. He showed that the force exerted on a planet by the sun must vary inversely to the square of the radius of the planet's orbit. (Mathematically, this is written as $1/r^2$.) From this relationship he developed his theory of gravity and his formula for determining gravitation.

In 1668 his interests in optics and the planets converged.

Newton knew that the lenses used to focus light in telescopes of his time (which we now call **refracting telescopes**) inevitably caused some distortions in the images produced. Different colors of light are bent at slightly different angles as they pass through the lenses of refracting telescopes—something Newton himself discovered with his prisms. He found a way around this distortion by inventing a new kind of telescope, one that used a curved mirror at the bottom of the telescope tube to focus the light on a flat mirror attached to an eyepiece near the tube's top. Such a telescope is called a **reflecting telescope**. The largest telescopes in the world, including the orbiting Hubble Space Telescope, are reflectors. Not only did Newton change the course of astronomy through scientific experiments with light and prisms, he also did so by toolmaking.

In 1667, when the Plague subsided and the ashes of London had cooled, Newton returned to Cambridge. He had left as a semiobscure graduate. Although his *Principia* had not yet been published, he returned as a world-renowned thinker. Two years later, at age twenty-seven, he was appointed to the prestigious Lucasian chair of mathematics at Cambridge. (Today that honor is held by another world-shaking thinker, Stephen Hawking.) In 1672 he was elected to the Royal Society, the most prestigious group of scientists and thinkers in the Western world.

Although most of Newton's work was known about and celebrated, it would not be published for nearly another twenty years. The reason was simple: the man was so terrified at the thought of his work being published that he simply didn't do it. Finally, in 1685, with the persistent encouragement of his friend Edmond Halley (1656–1742), Newton finally got around to writing down the details of his experiments and discoveries of two decades earlier. The *Philosophiae Naturalis Principia Mathematica* was his masterpiece and remains to this day one of the greatest scientific works of all time. The *Principia Mathematica* was published in its original Latin version in 1687, mostly through the work and funds of Halley. An English-language version was published in 1729, two years after Newton's death.

In the *Principia Mathematica* Newton laid out the famous mathematical principles, axioms, and laws that would govern physics for the next two hundred years. He stated, for example, his belief in the flow of "absolute, true, and mathematical time" and in the immutability of "absolute space, in its own nature, without relation to anything external." He defined "absolute motion" as the movement "of a body from one absolute place to another." This belief in absolute space, time, place, and motion was a remnant of Aristotle's philosophy. Newton himself must have known even then that there is really no such thing as "absolute space" or "absolute time." Even in the seventeenth century it was obvious that differently moving observers would assign different coordinates to the same "absolute" point in space. However, Newton resolved the problem in a way that is familiar to many even today: He simply appealed to God. Absolute space, he said, is absolute in relation to God. Thus, to deny absolute space would be to deny the existence of God—a dangerous thing to do in 1687.

Newton also wrote about his discovery of the relationship between force and motion, as stated in his famous three laws of motion:

1. A body stays in a state of rest or in motion at a constant speed in a straight line, unless its state is changed by forces from the outside—the principle of inertia.

2. The rate of change in a body's motion is inversely proportional to its mass and proportional to the amount of force applied—the principle of acceleration.

3. For every action, there is an equal and opposite reaction—implying the principle of conservation of a system's total momentum.

Finally, Newton in the *Principia Mathematica* expounded on his theory of gravity. He combined his work on force and motion with Kepler's laws of planetary motion to arrive at a law of universal gravitation. He showed that a central force causes objects to travel elliptical orbits (Kepler's first law), and that objects that travel under the influence of this force obey Kepler's second law of areas. From the geometric properties of ellipses, Newton demonstrated that the central force causing elliptical orbits can be described by an inverse-square law. With that in hand, he was able to mathematically derive Kepler's third law of harmonics. Thus, he showed that his discoveries fit the laws of planetary motion as they were known during his time.

Contrary to the popular story, there was probably no apple falling from a tree and hitting Newton on the head. However, Newton himself remarked that his observation of falling objects such as apples inspired him. He made what was then an enormous intuitive leap. If an apple's fall to Earth is caused by gravity, could it be that the moon stays in its orbit around the Earth because of the same force? "I began to think of gravity extending to the orb of the moon," he wrote, thus making the jump from a local force to a universal one. He then showed that the force of gravity predicted solely from his calculations is in close agreement with the gravitational force actually observed to be keeping the moon in its orbit. (Newton's universal law of gravitation, written in modern-day algebraic form, is $F = GMm/R^2$. F is the gravitational force between two bodies with masses M and m, whose centers are separated by the distance R. G is a constant called the gravitational constant.)

The other major book by Newton was his *Optiks*, published in 1702. Here he explained his experiments and speculations about light—its colors, reflections, refractions, and the nature of the "corpuscles" (we call them photons) of light.

Newton's discovery of the universal gravitational force had immense consequences for science in particular and for humanity's cosmological vision in general:

- Newton's gravitational force explained the physical interaction of the sun and planets, something that Kepler had vainly sought to uncover. Thus, it confirmed and strengthened the validity of Kepler's three laws of planetary motion.

- Gravity as explained by Newton also made it clear why objects do not fly off an Earth that Copernicus (and before him Aristarchus) asserted was rotating on its axis once a day.

- The predictive power of the Newtonian explanation of gravity was dramatically proved by one of his closest friends.

From ancient times people had believed that comets (the word
comes from the Greek *kometēs*, meaning "long-haired") were
some mysterious atmospheric phenomenon. Edmond Halley
showed that comets orbit the sun under the influence of
gravity. In particular, Halley discovered that the bright comets
that had appeared in the sky in 1531, 1607, and 1682 all had
nearly identical orbits. He correctly deduced that these three
comets were actually the same comet, returning every seventy-
five to seventy-six years. Halley predicted the appearance of
the same comet in the year 1758. True to his prediction, a
German farmer spotted the comet in December of that year.
Edmond Halley was by then fourteen years in his grave. To
this day, however, that comet—Halley's comet—bears the
name not of its discoverer but of the man who used Newton's
theory of gravity to predict its return.

• Newton's gravitational theory was also used to discover
a new planet (Neptune) and a new class of stars called binary
or double stars. In 1845 the British scientist John C. Adams
used Newton's equations to explain the deviation of the planet
Uranus from its predicted orbit. Adams theorized that a large
unknown planet beyond Uranus was affecting Uranus's orbit
with its gravity. He predicted where the planet should be in
the sky and told the Astronomer Royal, Sir George Airy, of his
calculations. Airy ignored them.

At the same time, the Frenchman Urbain J. J. Leverrier had
carried out calculations similar to those of Adams. Leverrier
passed his prediction on to astronomer Johann Galle in Berlin.
On September 23, 1846, Galle pointed his telescope at the
place in the sky Leverrier said to look. And there it was: a new
planet, later named Neptune.

Meanwhile, the discover of the planet Uranus also
uncovered the existence of binary or double stars by applying
Newton's law of gravity. Throughout the latter years of the
eighteenth century William Herschel and his sister Caroline
had been observing many pairs of stars. They hoped to
observe the phenomenon of heliocentric parallax—the
apparent movement of nearby stars against the more distant
background, revealing their actual distance from the Earth.

The Network Nebula forms the northeastern edge of the Cygnus Loop, the vast supernova remnant in the constellation Cygnus. The Loop together with its various parts form an emission nebula. The gases that constitute the nebula are ionized and shine by their own light rather than reflected light from nearby stars. The Cygnus Loop lies about 2,600 light-years from Earth.
Lick Observatory photograph

The open Type Sc spiral M101 in the constellation Ursa Major
Lick Observatory photograph

47 Tucanae is one of hundreds of globular clusters that orbit the Milky Way Galaxy in the region known as the Galactic halo. Globular clusters contain anywhere from several hundred thousand to a million or more stars.
Anglo-Australian Observatory; photograph by David Malin

NGC 7293 is a giant planetary nebula, a shell of gas thrown off by a star during the final stages of its life.
Lick Observatory photograph

M81 in the constellation Ursa Major is a Type Sb galaxy. Its central bulge is rather larger and its spiral arms more tightly wound than those of an Sc galaxy. Our Galaxy probably looks much like M81. Note the spurs coming off the upper spiral arm.

National Optical Astronomy Observatories

NGC 4565 is a spiral galaxy seen edge-on, revealing the dark lanes of gas and dust in its disk. Our own Galaxy would probably look like this if we could see it edge-on from the outside.

Lick Observatory photograph

M51 in the constellation Canes Venatici is a Type Sc spiral galaxy with open spiral arms and a small galactic bulge. M51 is notable for its unusual satellite galaxy, seen to the right of this image.
National Optical Astronomy Observatories

The globular cluster M92 in the constellation Hercules
Lick Observatory photograph

M32 is one of the satellite galaxies of the Andromeda Galaxy (M31). It is a Type E2 elliptical galaxy.
National Optical Astronomy Observatories

The Rosette Nebula is an H_{II} region about 4,500 light-years from Earth. It is heated and ionized by a group of young stars lying near its center. This close-up of the Rosette's northeastern quadrant shows dark filaments and globules silhouetted against the nebulosity. The globules are probably regions of new star formation.
Lick Observatory photograph

The Sagittarius region of the Milky Way. This region of the Milky Way includes the Galactic center, but it is obscured from naked-eye view by the dense, dark clouds of gas and dust.

Photograph courtesy of the California Institute of Technology

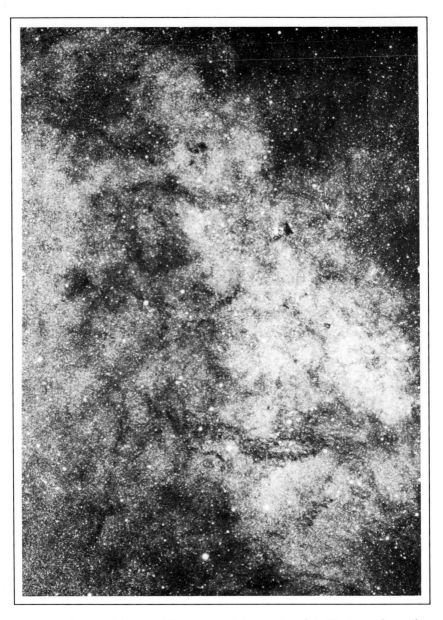

This is a close-up of the Milky Way star clouds in a region of Sagittarius to the south of the Galactic center. The dark areas are not regions with fewer stars, but rather clouds of opaque dust and gas that block our view of the stars beyond.
Lick Observatory photograph

What they actually did see was what seemed to be the motion of some stars around one another. By applying Newton's gravitational equation, the Herschels showed that these stars were actually orbiting one another.

Other discoveries and breakthroughs by Newton have also had an incalculable impact on science. The planet Uranus was discovered (by William Herschel in 1781) with the use of a reflecting telescope, invented by Newton. The entire edifice of classical and quantum physics begins with Newton's discovery of the spectrum and his suggestion that light is made of tiny "corpuscles" we today call photons. And the invention of calculus by Newton and independently by Leibniz marks the beginning of modern mathematical science. If these triumphs were his entire legacy, Newton could rightly be proclaimed one of the two or three greatest scientific thinkers of all human history.

Yet even more important than all these triumphs is the fundamental "paradigm shift" that Newton caused. From the Greeks to Ptolemy and even through Copernicus, the cosmos was envisioned as a closed and finite system, its boundaries marked by the sphere of stars. Newton's discovery of universal gravitation, however, *demanded that the universe be infinite in extent.* As he himself noted, if the universe were finite, then the force of gravity would eventually cause all objects in the cosmos to fall together into one big lump. This was manifestly not the case. By contrast, matter in an infinite universe would be pulled by gravity into an infinite number of objects. Since there are clearly many objects in the universe—from stars to the planets, the sun, and the Earth itself—Newton concluded that the universe must be infinite. Thus, Newton's discovery of gravity, and of its universality, metaphorically caused the universe to expand from a tiny set of crystalline orbs into a sphere of infinite extent. The cosmos became, to paraphrase the ancient Hellenistic writer, a sphere "whose center is nowhere and whose circumference is everywhere." Alexander Pope did not exaggerate when he wrote:

> Nature and Nature's laws lay hid in night:
> God said, *Let Newton Be!* and all was light.

Newton's view of the cosmos was not perfect in every respect. He himself knew that his insistence on the existence of absolute space and time did not quite jibe with reality. He was also bothered by the mutual gravitational interaction of the planets. Newton was sure that this would eventually cause the solar sytem to collapse into chaos. He had to appeal to Divine intervention to resolve both these problems.

Another, more concrete difficulty had to do with the planet Mercury. The axis of its orbit itself rotates in space, a phenomenon called **precession**. Some of this could be explained by the gravitational pull of the other planets, but not all. The only reasonable explanation that astronomers could devise was the presence of another planet, closer to the sun than Mercury. This hypothetical planet was dubbed Vulcan (absolutely no connection to the hypothetical science fiction TV planet two centuries in the future), and its gravitational pull was supposed to explain the precession of Mercury's orbit. The Vulcan theory had one large hole: no one could find it. And it never has been found. It does not exist. The solution to the puzzle of Mercury's precession would have to wait until long after Newton's death.

Except for these few nagging problems, the Newtonian vision of the cosmos marched from triumph to triumph, its predictions validated again and again. Less than two centuries after his death, the cosmological vision of the brilliant, eccentric failed gentleman farmer from Lincolnshire had triumphed (at least in the Western world). It would take the efforts of an obscure young German working in a Swiss patent office to displace the worldview created by Isaac Newton.

5
FROM HERSCHEL TO BOHR

The next truly major change in our vision of the cosmos was carried out by a German war refugee who eventually was knighted and became a court astronomer of the King of England. The life story of William Herschel (1738–1822) would make a great TV miniseries, but what is most important for our journey to the center of the Galaxy is three of his astronomical breakthroughs: the discovery of the planet Uranus, binary stars, and the Milky Way Galaxy.

William Herschel and the Galaxy

First things first: on the night of March 13, 1781, fifty-four years after Newton's death, William Herschel discovered a new planet. It was a dramatic accomplishment in a year that saw many, including the conclusive triumph of the American colonial rebels over the British at Yorktown; the arrival of Franciscan missionaries in a tiny village that would later become known as Los Angeles; the publication of Kant's *Critique of Pure Reason*, Rousseau's *Confessions*, and Haydn's Austrian string quartets; and the beginning construction of the trans-Siberian highway.

By this time William Herschel (whose real name was Friedrich Wilhelm Herschel) had been living in England for twenty-four years. He had fled Germany in 1757 during the Seven Years' War and settled in the town of Bath. There he earned a good living as a musician and music teacher, and indulged his habit— perhaps obsession is a better word—of stargazing. He had become arguably the best builder of reflecting telescopes in all of

Europe and England. One of his scopes had a mirror about 15 centimeters in diameter (6.2 inches; parents today buy similar-sized telescopes as birthday or Christmas presents for their children) and a focal length of about 2.1 meters.

On the evening of March 13, 1781, Herschel was in the middle of a systematic search for stars that seemed to come in pairs—binary or double stars. He was using his 15-centimeter reflector with a 227-power eyepiece and looking at some stars in the constellation Gemini ("The Twins"). Quite unexpectedly, Herschel spotted a "star" that was visibly larger than the others. He knew that even when seen through powerful telescopes, stars show no visible disks. They are obviously at a great distance from the Earth. Yet this object definitely had a disk. Herschel decided it was a comet, much like the one whose return in 1758 Edmond Halley had correctly predicted.

Herschel was intrigued, and he kept a close watch on his new "comet." He also reported his observations to Neville Maskelyne, the Astronomer Royal. By the beginning of April it was clear that this was no ordinary comet. It had no tail, and the edges of its disk were sharp instead of fuzzy. By the end of the month Maskelyne was writing to Herschel to congratulate him on his discovery of "the present Comet, or planet, I don't know what to call it."

It was a planet, the first to be discovered since Herschel's distant ancestors had noticed the five "wandering stars" in the night sky. Its orbit was nearly circular, like that of the five classic planets, and the orbit lay far beyond that of Saturn. In fact, Herschel's new planet was twice as far from the sun as Saturn. By the end of the year William Herschel had become astronomy's first media superstar. King George III knighted him, granted him a permanent government salary, and made him an official court astronomer. Herschel and his sister Caroline (who would become a renowned astronomer in her own right) moved from Bath to Windsor. The Royal Astronomical Society awarded Herschel its prestigious Copley Medal for his discovery of—

—Of what? What was the new planet's name? Today we call it Uranus, a name suggested by several astronomers during Herschel's time. However, Herschel himself had wanted to name it "Georgium Sidus" (the Georgian Star) in honor of his royal patron; the German refugee was not at all embarrassed to kiss the hand that was feeding him. Most of his contemporaries preferred the name "Herschel" for the new planet, and in fact that's what it was frequently called for nearly a century.

What was important about this discovery was that Herschel had essentially doubled the size of the known solar system. Newton's discoveries had revolutionized humanity's cosmological vision of the universe, but that had been primarily philosophical and mathematical. Herschel's discovery of Uranus was something concrete. See, his discovery said, the cosmos really *is* bigger than we ever imagined. What's more, the discovery of Uranus once and for all destroyed the old cosmologies of five planets, crystal spheres, and a fixed starry dome. Those world visions had no room for new planets. Yet here was one. It was the end of a mythology and the beginning of the rise to prominence of the one we have today.

Herschel's second significant contribution to astronomy was the discovery of the nature of binary stars. As we just saw, Herschel discovered Uranus while engaged in a systematic search for double stars. The search was a means to an end. Like other astronomers of his time, Herschel wanted to determine the distance from the Earth to the stars. Of course, everyone knew by then that the Earth was not at the center of the solar system; the sun was. "Everyone" also "knew" that the sun was at the center of the universe. So it seemed like a reasonable thing to find out just how big Newton's "infinite" cosmos really was, by measuring the distance to some stars.

It appeared that the only way to do this was to make measurements of heliocentric parallax. This apparent movement of stars against the more fixed and unmoving stellar background would give some hint of their distance from the sun. For this purpose, Herschel decided to use double stars, stars that appear to lie extremely close together in the sky. In fact, it seemed that wherever he looked with his new and powerful reflecting telescopes (invented, you may recall, by Isaac Newton), he found that many apparently single stars were actually double—two stars lying so close together that to the naked eye they seemed one. This, then, was the reason why William and Caroline were carefully finding double stars, noting their positions in the night sky, cataloging them, and going on to look for more. Herschel's systematic observing techniques were rare for his time. In fact, he is often considered to be the first truly professional astronomer and the founder of today's systematic *science* of astronomy.

It was while carrying out their parallax studies that the Herschels unexpectedly discovered movement *by the stars in some of the pairs*. When William Herschel calculated the nature of these movements, it became clear that gravitational forces were at work. The stars in many of the pairs were not lying close together solely by coincidence. In two papers published in 1803 and 1804, Herschel conclusively demonstrated that the two stars that make up the visual double star Castor in the constellation Gemini are gravitationally bound together. They follow Kepler's laws of planetary movement and Newton's law of universal gravitation. Today such double stars are called binary stars or binaries.

The cosmos expanded again, thanks to Herschel. Newtonian gravity not only applies to the Earth and the moon, and the sun and the Earth, and the sun and the planets, and even the sun and Uranus. It also applies to two stars orbiting each other, stars so distant from the Earth that their parallax could not be detected.

Herschel's third important astronomical advance is quite pertinent to our journey into the Galactic core. For it was William Herschel who made the first scientific attempt to determine the shape of the cosmos. In the process he came surprisingly close to discovering the shape of the Milky Way Galaxy.

This took place because of his failure to find any parallax movements in his study of double stars. Herschel decided that he was going to need some other technique to determine the distribution of stars in the cosmos. Accordingly, in 1784 he devised the method he called "star-gauging." He pointed his telescope at specific areas of the sky and carefully counted ("gauged") the stars that he could see. In just one year Herschel gauged the stars in more than fourteen hundred selected spots in the sky—an incredible feat of astronomical observation! In 1785 he published the results of star counts made in 683 different regions (he later added the results of four hundred other regions he had also looked at but had thought not worth publishing). In some areas he saw only one star; in others he counted six hundred or more, and in one region he estimated he had seen *116,000* stars. And, yes, there was a method to this apparent madness of stellar counting.

It is obvious to anyone looking up at the night sky that the stars are distributed unequally. Some areas have more stars than others. Galileo had used his primitive telescope to discover that the Milky Way was actually composed of a myriad of faint stars, which only confirmed the evidence of the naked eye. This inequality of distribution could have two possible causes: it could be due to an unequal distribution of stars in three-dimensional space, or to a difference in the distance to which different parts of the stellar system extend. Suppose you are standing in a forest. The woods seem to be thicker in one direction than in another. This could be because the trees really are planted closer together in that direction. Alternatively, it could be an illusion caused by the fact that the forest simply extends further in that direction. That means there are more trees stretching off into the distance, so it *looks* as though a lot of them are crammed into one space.

Herschel decided that the concentrations of stars in different areas of the sky were a measure of the depth to which the stellar system extended in that direction. He interpreted the results of his star gauges accordingly. Herschel confirmed the evidence of the unaided eye, that stars are most plentiful in and around the Milky Way and most sparse in those regions of the sky farthest from it. The picture that emerged from his star counts was of a stellar system with the shape of a disk or a grindstone. According to his calculations, its diameter was about five times its thickness. One side extension of the grindstone was divided into two branches, and the space between them seemed almost completely free of stars. The sun and its retinue of planets were located near the center of the grindstone. The physical size of the stellar grindstone was unknown, since no one then had determined the actual distance to any star. But at least Herschel had come up with some kind of shape to the distribution of stars.

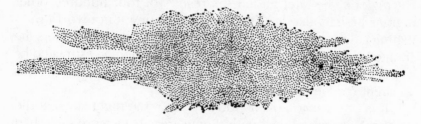

William Herschel's "grindstone" model of the Milky Way Galaxy had a diameter about five times its thickness, with two branches of stars extending out of one side and the sun located somewhere near the center.
Illustration from Arthur Berry, *A Short History of Astronomy: From Earliest Times Through the Nineteenth Century,* Dover Publications, Inc., 1961. Reprinted by permission of the publisher.

This system of stars, this stellar grindstone, was understood to be the universe—not the Galaxy as we call it today, but *the entire known universe*. In Herschel's time there was no knowledge of external galaxies, superclusters, cosmic redshifts, black holes, and quasars.

Today we know that Herschel's grindstone model was inaccurate at best. For one thing, the apparent density of stellar trees in the cosmic forest is not only due to the extent of the forest, but also to real concentrations of trees in some areas. There are many real concentrations of stars in the Galaxy, including things like open star clusters and globular clusters. Herschel himself admitted in 1811 that his assumptions were probably wrong to begin with and that his grindstone model of the universe was probably mistaken. For another thing, the apparent gap between two branches of the grindstone turns out to be just that—apparent. Today we know that many areas of the sky that appear to be empty of stars actually look that way because of the presence of vast interstellar clouds of gas and dust. These clouds block our view of equally vast expanses of stars, especially our view of the center of the Galaxy itself. Then there is the matter of the size of the universe. Herschel could not know, as we do, that the cosmos is more vast than in his wildest eighteenth-century dreams.

Curiously, though, the Herschel grindstone or disk model of the universe is not all that far from the actual shape of the Milky

Way Galaxy. The real thing is flatter than his version, and we're nowhere near the center—but it was a good guess.

James Clerk Maxwell

One of the great scientific minds of the nineteenth century, James Clerk Maxwell, was born in Edinburgh nine years after the death of William Herschel, and he died in Cambridge, England, in 1879. He graduated from Cambridge when he was only nineteen, and took his first job as a professor of mathematics at the University of Aberdeen in 1856. Maxwell's great mathematical discovery brought to fruition the work of the seventeenth-century Dutch astronomer Christian Huygens (1629–1695) and also set the stage for the dramatic arrival of Albert Einstein.

Curiously, Maxwell's first significant contribution to science was an explanation of another breakthrough by Huygens—the discovery of the rings of Saturn. Huygens, who was a contemporary of Newton, had in 1656 discovered the Orion Nebula, the moon of Saturn he named Titan, and also that planet's system of rings. Huygens thought the ring system was a solid structure. However, Maxwell showed that a solid structure could not possibly exist in orbit around Saturn without being torn apart by its differential rotation. Different parts of a solid ring would want to orbit Saturn at different velocities, and that would break it apart. Rather, said Maxwell, it is clear that the Saturnian ring must be made of small particles that orbit the planet independently.

However, Maxwell's greatest achievement was to provide convincing mathematical evidence that light can be described as a wave. This had been Huygens's suggestion back in the seventeenth century. Huygens believed that light is a form of vibration that travels in a straight line. Since this contradicted Newton's assertion that light is "corpuscular," made of particles, Huygens went looking for evidence to prove his position. He discovered that waves can also travel in straight lines, and they obey the laws of reflection and refraction, with which he—an expert maker of lenses—was familiar. He believed that light waves are propagated through a medium made of infinitesimal particles, a medium he referred to as the "ether." The waves created by the displacement of many of these ether particles create visible light.

Huygens's wave theory of light was greatly overshadowed by the Newtonian corpuscular theory. However, it was to steadily gain ground in physics. In 1802 the English physicist Thomas Young (1773–1829) announced his discovery that intersecting beams of light cause interference patterns. This is something that waves do—like overlapping waves in water caused by two pebbles dropped in a pond. Beams of particles do not create interference patterns. Clearly, light was a wave phenomenon.

The next step was taken by the brilliant self-taught scientist Michael Faraday (1791–1867), who in 1821 carried out experiments that showed a linkage between electricity, magnetism, and light. Ten years later—the same year Maxwell was born—Faraday realized that electrical fields and magnetic fields are different aspects of the same thing, an electromagnetic field. However, one of the things Faraday had not taught himself was mathematics. He was unable to put his insight into any concrete form. It was James Clerk Maxwell who did so, completing the work begun by Huygens more than two centuries earlier.

It was the year 1864. Tolstoy began writing *War and Peace*, and General Sherman was marching through Georgia. French painter Henri Toulouse-Lautrec was born, Nathaniel Hawthorne died, and Karl Marx founded the First International Workingman's Association. James Clerk Maxwell derived a set of four partial differential equations that described the behavior of electrical and magnetic fields. They showed mathematically (something Faraday could not do) how electrical and magnetic fields are related to each other. A changing magnetic field creates a change in its associated electric field, and vice versa. The field expands outward in all directions, and *it moves outward at the speed of light*. Heinrich Hertz would later use this discovery to prove that light itself is part of the electromagnetic spectrum.

Maxwell's equations were the crowning touch to the growing scientific conviction that light is made of waves, not particles. Maxwell himself was convinced of it. He, like Huygens, concluded that some invisible medium through which electromagnetic waves travel must permeate all of space.

Waves of Energy

Electromagnetic energy moves through space in the form of **electromagnetic waves**. These waves are somewhat similar to the waves we see on the open ocean or the waves that can be created in a length of rope by snapping it up and down or back and forth. One important difference is that ocean waves and the waves in a vibrating string or rope travel through some kind of material medium—the ocean or the string. Electromagnetic waves do not need some material medium in which to travel. In 1887 physicists Edward Morley and Albert Michelson conclusively proved that the ancient (and not so ancient) idea of a space-filling "ether" in which light waves travel is wrong. There is no such thing. Nevertheless, electromagnetic waves do exist, and they do travel through space. These waves of course have wavelengths, which refer to the length from the crest of one wave to the crest of the next. There is a great variation in wavelength for different kinds of electromagnetic waves. The wavelength of AM radio waves, for example, is around three hundred meters, while the wavelength of some cosmic rays is measured in femtometers, or quadrillionths of a meter.

Electromagnetic waves come in many different frequencies. **Frequency** is a measurement of the number of times a wave moves through its complete cycle in some unit of time. A single complete cycle for a wave starts at some equilibrium or neutral point, moves to a maximum positive variation, back to equilibrium, then to a maximum negative variation, and back to the equilibrium point. The most common measurement of frequency is the **hertz** (abbreviated as Hz). (This unit of measure is named for Heinrich Hertz, the nineteenth-century German physicist who studied radio waves.) Some kinds of direct electrical current have a frequency of less than 1 Hz, while scientists have detected gamma rays with a frequency of 10^{23} Hz. That's a hundred billion trillion, or the number 1 followed by twenty-three zeros.

The relationship between wavelength and frequency is an inverse one. The lower the frequency, the longer the wavelength, and vice versa. So AM radio waves have a low frequency but a long wavelength. X-rays, on the other hand, have a very high frequency, but their wavelength is quite short.

THE ELECTROMAGNETIC SPECTRUM

Frequency (Hz)	Wavelength (m)	Example
10^{23}	3×10^{-15}	Cosmic rays
10^{22}–10^{21}	3×10^{-14}–3×10^{-13}	Gamma rays
10^{21}–10^{20}	3×10^{-13}–3×10^{-12}	Hard x-rays
	2.43×10^{-12}	Electron-positron annihilation
10^{19}–10^{18}	3×10^{-11}–3×10^{-10}	Soft x-rays
10^{18}–10^{15}	3×10^{-10}–3×10^{-7}	Ultraviolet light
10^{15}–10^{14}	3×10^{-7}–3×10^{-6}	Visible light
10^{14}–10^{12}	3×10^{-6}–3×10^{-4}	Infrared, far-infrared
10^{11}–10^{10}	3×10^{-3}–3×10^{-2}	Microwaves
10^{10}–10^{9}	3×10^{-2}–3×10^{-1}	Radar
	0.21	Interstellar hydrogen
10^{8}	3.0	Television, FM radio
10^{7}	30.0	Shortwave radio
10^{6}	300.0	AM radio
10^{5}	3,000.0	Long-wave radio

An arrangement or display of electromagnetic radiation according to wavelength or frequency is called a **spectrum**. Radio waves, visible light, x-rays, and gamma rays are all part of the electromagnetic spectrum. Electromagnetic wavelengths and frequencies are somewhat more understandable when we look at the location on the electromagnetic spectrum of some familiar examples. AM radio in the United States has a typical frequency range of 550 to 1700 kilohertz (kHz). Shortwave radio has a frequency of around 10 megahertz (MHz) and a wavelength of about 30 meters. The frequency range of FM radio is from 88 to 108 MHz. The wavelength of FM signals is about three meters, one-tenth that of shortwave radio.

Visible light is on this spectrum, too, but at much smaller wavelengths and higher frequencies. And it occupies only a very small part of the entire electromagnetic spectrum. All of visible light, from dark red through deep violet, falls in a narrow band between 10^{-6} and 10^{-7} meters (1–30 micrometers, or microns, abbreviated μm). The frequency of visible light is around 10^{14}–10^{15} Hz.

Up at the far end of the electromagnetic spectrum are x-rays and gamma rays, with frequencies ranging from 10^{18} to 10^{23} Hz and wavelengths from 30 nm (nanometers, thousand-millionths of a meter) to 300 pm (picometers, a million-millionth of a meter).

Most of the objects in the universe cannot be seen very well in visible wavelengths of light. However, they do emit other wavelengths. To be more specific, objects emit electromagnetic radiation at wavelengths that depend on their temperature. The higher the temperature, the shorter the wavelength and the greater the frequency of the radiation emitted. Different atoms, molecules, and nuclear processes emit characteristic wavelengths of light. These wavelengths act as fingerprints for those atoms or molecules or energy-releasing processes.

Astronomers have learned to read those fingerprints by detecting these other frequencies of electromagnetic radiation and converting them into maps, images, or other pictures that are visible. For example, interstellar hydrogen atoms radiate energy at a wavelength of about 0.21 meters, or 21 centimeters. When astronomers detect radiation of this wavelength coming from some point in the heavens, they know it signals the presence of atomic interstellar hydrogen. Another example is the mutual annihilation of matter and antimatter electrons. Antimatter is matter made of antiparticles, which in turn are subatomic particles similar to protons, electrons, neutrons, and other normal subatomic particles. They are identical in some ways to ordinary subatomic particles but with an opposite electrical charge and spin. So the antimatter version of an electron (which has a negative electrical charge) is called a positron. It is the same size and mass as a normal electron, but it has a positive electrical charge. When matter and antimatter meet, they annihilate each other and turn into pure energy. In the case of positron-electron annihilation, that energy is at a unique wavelength of 2.43×10^{-12} m. Astronomers have seen this particular signal coming from near the center of the Galaxy, so they know that there is at least some antimatter in the form of positrons near the Galactic core.

The Key to Stellar Knowledge

Scientists in most other disciplines usually have something they can study firsthand—plant or animal cells, strands of DNA, ancient bones, shards of prehistoric pottery, water samples, pieces of rock. With the exception of those who study meteorites, comets, and the moon, astronomers have only one "thing" they can study—light. We have yet to gather a sample of Sirius, or a piece of a neutron star, or a cupful of gas from a giant interstellar molecular cloud. Astronomers can only observe the light that is either emitted by or reflected by such objects.

However, light can tell us a lot, if we know what to look for. Studying the wavelengths and frequencies of light from astronomical objects gives us some information about those objects. Another way to study light is to break a beam of light up into its component colors. This is called **spectroscopy**.

Nearly everyone has seen at least one form of an electromagnetic spectrum, probably many times. This type of spectrum is called a rainbow and is naturally produced. It is quite easy to create your own rainbow or visible-light spectrum at home. You need the triangularly cut piece of solid glass called a prism. When a beam of white sunlight is passed through the prism, the different wavelengths of visible light are bent (the technical term is **refracted**) at different angles. Red light is bent the least, and violet light the most. The result is that the "white" light is broken into its constituent colors. Those colors are spread out across the table, chair, wall, or whatever lies on the other side of the prism. You see not a beam of white light, but a rainbow display of the light's true colors. The spectrum is a band of colors ranging from red at the longest wavelength and lowest frequency, through orange, yellow, green, and blue to indigo and violet at the shortest wavelength and highest frequency.

Another way to create a spectrum is with a **diffraction grating**. A diffraction grating is an arrangement of narrow slits or grooves that **diffract** or bend different wavelengths of light at different angles. By controlling the shape and size of the grooves when making a grating, and by shining the light on the grating at different angles, you can create spectra with brightness and purity much greater than can be achieved with a prism. Diffraction gratings are now routinely used instead of prisms in building astronomical instruments such as spectroscopes and spectrographs.

Astronomical spectroscopy is one of the most powerful techniques for learning about the composition, physical conditions, and velocities of astronomical objects. It is possible to gather this kind of information because of the appearance of different kinds of lines in astronomical spectra. When solids or liquids, or gases under high pressure, are heated to incandescent temperatures, they produce a **continuous spectrum** that contains all the colors. When a gas at low pressure is heated to incandescence, its spectrum is a series of separate glowing lines

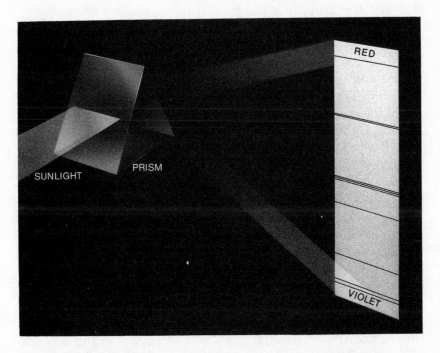

When a narrow beam of sunlight passes through a prism it is broken up into the component colors of the spectrum, from red through violet. What's more, the spectrum of sunlight contains a series of dark lines called Fraunhofer lines.

Illustration from Jay Pasachoff, *Contemporary Astronomy*, third edition, Saunders College Publishing, a division of Holt, Rinehart and Winston, Inc., 1985. Reprinted by permission of the publisher.

of light called an **emission spectrum**. These glowing lines may sometimes be superimposed on a continuous spectrum. When light from a continuous spectrum passes through cooler matter, like a cloud of cool gas, the continuous spectrum will have superimposed on it a series of *dark* lines. This kind of spectrum is called an **absorption spectrum**. These three categories were formulated by Gustav Kirchhoff and are called Kirchhoff's rules for spectral analysis. Any spectrum that has discrete spectral lines, be they absorption or emission lines, is generally referred to as a **line spectrum**.

The lines in emission and absorption spectra are not randomly placed. They appear for a specific reason. One way to understand why and how is to look at the emission spectrum of hydrogen. The low-pressure hydrogen gas emitting the light and thus the spectrum first has to absorb some kind of energy to heat it up to incandescence. That energy raises the hydrogen atoms to higher energy levels. A hydrogen atom is a simple creature, the simplest atom in the universe. It is made of one proton and one electron. The electron is commonly pictured as "circling" the proton as a planet circles the sun. This isn't really the case (it is closer to the truth to say that the electron is sort of everywhere around the proton as long as you don't look too closely), but it will suffice for the moment. All things being equal, the electron has a preferred "orbit," its location of lowest energy. There are also other locations around the proton that it can occupy. These are higher energy levels to which it moves only if enough energy is poured into the atom. When such an energy level is reached, the electron instantly "jumps" to that higher level. The hydrogen atom is now in what is called an "excited state."

However, the electron doesn't stay there very long. It quickly jumps back down to a lower energy level. As it does so, it releases discrete packets (or quanta—thus the phrases "quantum jump" and "quantum physics") of energy in the form of photons, which are the particles of electromagnetic energy. The emission spectrum of hydrogen or any other element is formed by these emissions of photons. The photons have characteristic frequencies depending on the atom that emitted them and the energy level from which they were emitted. Thus, the emission spectrum of hydrogen (or anything else) consists of a specific pattern of **emission lines** that occur at specific frequencies. The lines are called emission lines because the electron *emits* photon energy as it jumps down. (And **absorption lines** are so named because the electron *absorbs* energy as it jumps up one or more energy levels.)

Hydrogen's emission and absorption spectra are relatively simple compared to those of other elements. They are also extremely valuable to astronomers, because hydrogen is by far the most common element in the universe. Atomic hydrogen (usually referred to by astronomers as H_I; the letter H stands for hydrogen, and the roman numeral I means the atom's electron is present) has a line spectrum with several series of lines. One of these sets of lines, which is mostly found in the visible part of the hydrogen spectrum, is called the **Balmer series**. It is named for the Swiss physicist Johann Balmer, who in 1885 showed that the lines form a series that can be represented by a particular mathematical equation. The Balmer series of lines is produced when an electron jumps either from or to its second energy level or orbit around the proton nucleus.

For example, when the electron jumps from its second to third energy level, it causes a specific absorption line located at a specific place in the spectrum. When it jumps from its third level back down to the second level, it creates an emission line in that same location. This particular spectral line is one of the most famous and most important in astronomy and physics. It is called the **hydrogen-alpha** or Hα line. It falls at a wavelength

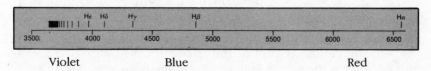

As the electron in a hydrogen atom jumps to and from different energy levels (or orbits) around the nucleus, it releases or absorbs energy at distinct wavelengths. This gives rise to different sets of lines in the hydrogen spectrum. The Balmer series represents transitions to or from the second energy state of hydrogen. The strongest line in the series, hydrogen-alpha, is in the red part of the spectrum.

Illustration from Jay Pasachoff, *Contemporary Astronomy*, third edition, Saunders College Publishing, a division of Holt, Rinehart and Winston, Inc., 1985. Reprinted by permission of the publisher.

of 656.3 nm (a nanometer is a thousand-millionth of a meter), which is in the red part of the visible spectrum.

The second line in the Balmer series is not surprisingly called hydrogen-beta (Hβ). It falls at 486.1 nm, in the green-blue end of the spectrum. As with the Hα line, emission lines are created when the electron jumps down from a higher level to the second level; absorption lines occur when the electron jumps from its second energy level to one of the higher levels.

Another famous series of lines in the hydrogen spectrum is the **Lyman series**. It is formed by transitions to or from the first or ground level of the electron. The Lyman series is located in the far ultraviolet region of the hydrogen spectrum. The most famous and important of its lines is the **Lyman-alpha** or Lα line, at 121.6 nm. Another set of lines, called the **Paschen series**, is caused by electron transitions to and from the third orbit. These lines are found in the near-infrared region of the spectrum. A series called the **Bracket series** involves hydrogen's fourth electron orbit, and its lines occur in the far-infrared region.

The importance of these and other lines in an astronomical spectrum is severalfold. First, emission or absorption lines are like fingerprints. *Each element has its own unique set of spectral lines, which act as unambiguous fingerprints for its presence.* This is relatively easy to see in the laboratory. If you burn a flame of hydrogen gas and look at it through a **spectroscope** (an instrument that uses a prism or a diffraction grating to make and observe spectra), you will see four bright lines. One is in the red end of the spectrum, one is in the green-blue region, and two are in the violet region. Now burn a sample of sodium in a hot, colorless flame. The flame will turn a bright yellow. Through the spectrograph, however, you will see two bright lines in the yellow region of the spectrum. You will also see a pair of lines in the green region, two in the orange region not far from the yellow pair, and a single line in the upper end of the red region. The patterns of these two sets of lines are completely different. So are the patterns of the emission lines from each and every element.

If a beam of white light from the sun is passed through the

burning sodium flame, you will see a pair of dark lines super-imposed on the continuous spectrum. These two lines are in the identical location of the pair of emission lines in the spectrum from the sodium flame. They are caused by the sodium in the flame absorbing the yellow light from the solar spectrum. If an astronomer sees a spectral line at 121.6 nm, the Lyman-alpha line, the astronomer *knows* that he or she is looking at hydrogen. Another line at another location would signal the presence of helium, or oxygen, or carbon.

One of the great astronomical breakthroughs in this realm of elemental fingerprints was by a German astronomer named Joseph von Fraunhofer. Beginning in 1814 Fraunhofer began a detailed study of the absorption lines in the spectrum of the sun's photosphere. The photosphere is the "visible surface" of the sun. It is the source of the absorption spectra of both the sun and most other stars. Fraunhofer eventually cataloged more than five hundred lines and labeled the most striking ones with letters—D lines, H lines, K lines, and so on. By following Kirchhoff's rules for spectral analysis, astronomers have been able to identify specific sets of Fraunhofer lines with specific elements. Fraunhofer's D lines are actually the doublet of lines at 589.0 and 589.6 nm, which are unique to sodium. The H and K lines are lines in the spectrum of singly ionized calcium at 393.4 nm (the H line) and 396.8 nm (the K line).

Today more than twenty-five thousand Fraunhofer lines have been identified in the spectrum of the sun. The ones most prominent in the visible part of the spectrum are mostly due to different forms of calcium, hydrogen, magnesium, and sodium. Many other lines are due to forms of iron.

The strength—that is, the width and darkness—of a Fraunhofer line is due not only to how much of a particular element is present, but also to its degree of ionization and the level of its energetic excitation. An element is said to be **ionized** when it has more or fewer electrons than is normal. An atom that is missing electrons is said to be positively ionized; it carries a net positive electrical charge because of the positively charged protons in its nucleus. If an atom has more than its usual complement of electrons, it is said to be negatively charged, because the excess electrons give the atom a net negative electrical charge. So, for example, when an astronomer says that he or she has spectrographic evidence of Ca+ in some cloud of interstellar gas, it means the astronomer has found evidence of the presence of singly ionized calcium atoms—calcium atoms missing one of their electrons. That evidence, as we've just seen, is the unique signature of the H and K Fraunhofer lines.

Whether the lines are absorption or emission lines also tells the astronomer something important. Emission lines, as we've seen, are created when a gas at low pressure is heated so much it glows. Suppose our astronomer is looking at the spectrum of a giant star and sees emission lines that follow the Balmer series. The astronomer may surmise that the radiation is coming from a shell of hot, diffuse hydrogen gas surrounding a star. Perhaps the star has recently exploded as a supernova, blowing off this shell. Or perhaps it is a star like Betelgeuse, the brilliant red supergiant star that makes up the shoulder of the constellation Orion. Betelgeuse hasn't blown up (not yet, anyway . . .), but it does periodically "puff away" an outer layer of its atmosphere. If the astronomer sees absorption lines, however, he or she knows that the source is radiation from a continuous spectrum passing through cooler matter. In this case, perhaps the radiation is coming from stars and has passed through an intervening cloud of cooler gas.

Still another clue comes from changes or shifts in the location of the spectral lines. These kinds of shifts are caused by the **Doppler effect**. The Doppler effect is a change in the apparent frequency and wavelength of an electromagnetic wave caused by the relative motion of the source and observer of the electromagnetic radiation. The effect is named for Austrian physicist C. J. Doppler, who first suggested it in 1842.

The degree of this change, when it is the result of radiation emitted by a moving object, is called the **Doppler shift**. When the object emitting the radiation is moving away from the observer, the wavelength is longer—and therefore redder—than it would be were the source and the observer stationary relative to one another. This shift to a longer wavelength is naturally called a **redshift**. When the emitting object is moving toward the observer, the wavelength is shorter than normal and is shifted toward the blue end of the spectrum. This is called a **blueshift**.

Redshifts and blueshifts make themselves apparent in the positions of spectral lines. By identifying the direction and degree of the shift, an astronomer can learn a lot about the object observed. Suppose an astronomer is recording the spectrum of a nearby star. The astronomer sees blueshifts in one part of the spectrum and redshifts in another. Additional observations may show that these shifts are caused by the rotation of the star in question. One side of the star is moving toward the observer (causing blueshifts), and the other side is moving away (causing redshifts). Careful measurements will reveal the star's rotational velocity—its "day."

Or suppose an astronomer detects an absorption line in the red part of a galaxy's spectrum that matches up with none of the elements. The astronomer then determines that it is actually the Lyman-alpha line but shifted far into the red part of the spectrum. This may be because the galaxy is moving away from the observer. The degree of the shift will tell the astronomer how fast the galaxy is moving away, or its radial velocity. Observations of the redshifts and blueshifts in spectral lines also help astronomers determine the orbital motions of binary stars (two stars orbiting around each other) and the relative movements of dust and gas clouds.

In fact, by observing redshifts and blueshifts in the spectral lines of gas clouds, stars, and galaxies, astronomers have utterly revolutionized our picture of the center of the Galaxy and of the very size and possible fate of the entire universe.

Michelson, Morley, and the Speed of Light

Despite these important applications, there were problems with the ether idea and the whole concept of light as a wave. While there was plenty of evidence that light acts like a wave, evidence also existed that light acts as though it is made of particles. The whole conundrum would not be resolved until the coming of Planck and Einstein. Meanwhile, the ether hypothesis came up for a rigorous test, which was administered by Edward Morley and Albert Michelson (1852–1931).

Born in Prussia, Michelson later came to the United States, where he eventually became head of the physics department at the University of Chicago. Michelson invented an instrument called the **interferometer**, which uses the interference patterns created by intersecting light waves to measure the speed of light to an extraordinary degree of accuracy. In 1887, at what is now Case Reserve Western University, Michelson and his colleague Edward Morley set up an experiment to determine variations in the speed of light caused by the movement of the Earth through the electromagnetic ether.

The assumption was a simple one, a matter of common sense. Suppose we are in a train traveling down the track at 25 kilometers per hour. Heading toward us on the same track is another train traveling at 100 kilometers per hour. Our combined closing speed is 125 kilometers per hour. Now suppose that the other train is heading away from us at 100 kilometers per hour. If we measure the velocity at which the two trains are moving away from each other, we will find it is 75 kilometers per hour.

Suppose, now, that light waves do indeed propagate through space by means of some invisible, massless substance called ether. It stands to reason that in the direction of the Earth's passage through space (and the ether), the speed of light would be c (the speed of light, which at that time had been measured at about 186,000 miles per second) plus some additional amount. The speed of light in the opposite direction will be c minus some amount. The same thing is true of the speed of sound waves, which propagate through air. It should be that way with waves propagating through the ether.

Michelson and Morley carried out several runs with their interferometer. They measured the speed of light to unprecedented accuracy. *And they found no difference whatsoever in the speed of light, in any direction measured.* The possible conclusions were inescapable and overwhelming:

• Either the ether does not exist, so light cannot be a wave—which contradicted the clear observations of many scientists;

• Or the ether does exist—but the Earth *contracts* in the direction of its motion through the ether just enough to cancel out the effects on the speed of light.

Both conclusions were utterly unpalatable to physics. But the Michelson-Morley experiments were clear and undeniable. Physicists and astronomers continued to believe that light behaves like a particle as well as like a wave. But they could not explain it.

Planck and the Ultraviolet Catastrophe

A satisfactory explanation for *why* absorption and emission lines exist in the spectrum was not easy to develop. According to the standard theory of physics in the nineteenth century, when atoms are excited they should emit light at a continuous range of frequencies, not at discrete frequencies that differ for each element. Yet there was no doubt that atoms were doing exactly what theory said they should not.

Science is based on observation of how the world works. If a theory makes predictions that do not match observations, then the theory must be changed. The changes that were to take place were monumental. They began at the end of the nineteenth century with the solution to the ultraviolet catastrophe.

At the end of the nineteenth century, physicists were faced with what seemed to be the total breakdown of their scientific framework of understanding. One excellent example of the failure of classical physics was the case of blackbody radiation. In physics, a blackbody is an ideal object that absorbs all of the radiant energy falling upon it and that reflects or emits none of that energy. A blackbody is also a perfect *emitter* of any radiation it might produce. Of course, there is no such thing as an actual *perfect* blackbody; it is a hypothetical ideal. Any object that has a temperature greater than absolute zero—and that's everything in the universe—will emit some radiation.

Some substances do come close to being perfect blackbodies. Lampblack, for example, reflects less than 2 percent of all incoming radiation. More important, the radiation that the sun and stars emit, as well as their temperatures, can be described by assuming that they are blackbodies.

The thermal or heat radiation that would be emitted by a blackbody at a particular temperature is called (not surprisingly) **blackbody radiation**. At lower temperatures the blackbody radiation is at longer (that is, redder) wavelengths, mostly in the infrared. The hotter a blackbody is, the more thermal radiation it emits at shorter wavelengths. The amount of radiation a blackbody emits is directly proportional to the fourth power of its temperature. In other words, if blackbody A is twice as hot as blackbody B, then A will emit *sixteen times* as much energy (2^4) overall as B. Also, each unit of A's surface, such as a square meter or a square foot, emits sixteen times as much energy as B, the blackbody that is half as hot.

The power radiated from a heated object depends on the radiation's wavelength, so it is simple in essence to plot a curve that shows the relationship between power and wavelength for any such object. This is called a radiation curve, and one can be plotted for a blackbody. The problem at the end of the nineteenth century was that physicists could not come up with a mathematical formula or law that would predict the observed curve for near-blackbodies. All the tools and theories of nineteenth-century classical physics failed to explain blackbody radiation.

The most famous attempt was called the Rayleigh-Jeans law. It worked well at long wavelengths (near the red end of the spectrum) but failed utterly at the shorter (ultraviolet) wavelengths. Scientists at the time referred to this as the "ultraviolet catastrophe," and they used the word *catastrophe* deliberately. It was just that.

In 1900 a German physicist named Max Planck came to the rescue. Perhaps *rescue* is the wrong word, however. Planck's solution to the ultraviolet catastrophe was more of a desperate leap into the dark than a rescue. Planck wrestled with the blackbody problem until he was mentally exhausted. He tried every approach, every trick, every wild idea he could think of to explain the nature of blackbody radiation from the point of view of classical physics. Nothing worked. Finally, in desperation, Planck simply decided *to change the rules*.

What Planck did was make a revolutionary assumption, which he combined with classical physics to develop a new formula for blackbody radiation. That assumption was that an object can emit radiation only in discrete amounts. He called such an amount or packet of energy a **quantum**. The word *quantum* was not new; it had first appeared in 1567 and meant "quantity" or "amount." Planck's use of it in physics, however, was radical. It had been an axiom of that science that energy *flows continually*. There are no breaks in it. It is like water. Planck was suggesting that energy is not something continuous, but rather that it comes in small units. It is discontinuous, like a stream of ball bearings.

As part of his explanation, Planck also invented a mathematical constant to use in his formula. He made up the constant (abbreviated h) because he needed it to make the mathematics work. When the math worked, it explained the blackbody effect. That's all. **Planck's constant**, as it is now known, was simply an expedient solution to Planck's unsolvable problem.

What Planck set out to do was resolve the ultraviolet catastrophe. What he actually did was overturn classical physics. And he hated it. To the end of his days, Max Planck fought against the validity of the creature spawned by his creation of the quantum—quantum mechanics. His stepchild was irresistible, though, and today reigns as the most predictively accurate theory of physics ever devised. The consequences of the quantum were indeed astonishing. But before quantum mechanics came an explanation of the **photoelectric effect**.

In 1905 Albert Einstein published three papers that shook the world of physics. One was his special theory of relativity (about which we will hear more anon), the second was an explanation of what is called Brownian movement, and the third was an explanation of the photoelectric effect. Discovered in 1887 by Heinrich Hertz, the photoelectric effect is the emission of electrons by substances such as certain metals when light falls on their surfaces. When the frequency of the light passes a certain threshold, electrons are emitted. Above that frequency, increasing the light's intensity increases the number of electrons released, but not their kinetic energy (energy of movement).

The puzzle of the photoelectric effect for nineteenth-century physicists was that reality did not match theory. Physicists believed that light is a wave. Were this the case, then there should be no such thing as a "threshold frequency." The electrons in the metal would continually absorb energy. At any intensity it would only be a matter of time before the electrons had absorbed enough energy to escape. The reality, though, was that below the threshold frequency, the light's intensity was irrelevant. It could be as intense as science could make it—but if the frequency weren't high enough, there would be no release of electrons from the metal.

Einstein explained the photoelectric effect by borrowing Planck's idea of the quantum. The light striking the plate, he said, carries energy in discrete amounts—in quanta. Each light quantum, or photon, contains an amount of energy that depends only on frequency, not intensity. Intensity depends on the number of photons in a light beam. The more photons, the greater the intensity. But the energy of each photon is determined by the frequency of the light. And the energy in each photon can be calculated by multiplying its frequency by the constant (h) that Planck had conjured up a few years earlier to help explain the problem of blackbody radiation.

Photons behave like particles that travel at the speed of light, said Einstein. And sometimes they behave like waves traveling at the speed of light. Photons are pretty strange creatures, indeed.

Bohr and the Quantum Atom

Planck's discovery that photons carry energy in the form of quanta, when coupled with other revelations about the nature of atoms themselves, eventually led to the explanation for the existence of emission and absorption lines in stars. Without these breakthroughs, astronomers would not be able to understand the cataclysmic events happening in the center of our Galaxy.

In 1897 the English physicist J. J. Thomson discovered the existence of a particle smaller than an atom and possessed of a negative electrical charge. It was named the **electron**. Electrons were responsible for the electrically negative currents that

Thomson produced when he immersed electrodes with opposite charges in a gas-filled tube. The same particles appeared, no matter what kind of gas or electrode material Thomson used. He concluded that electrons were a fundamental component of all matter.

In 1906 Thomson proposed that atoms consisted of electrons embedded in a positively charged background. The background could be likened to a thick pudding, with the electrons as raisins. The positive and negative charges of the pudding and the raisins canceled each other out, leaving the observed electrically neutral atom.

Thomson's model was scientifically testable. One need only bombard materials with beams of charged particles and see how those particles are deflected. The challenge was taken up by Ernest Rutherford, who was the leading expert in the then new and exciting field of radioactivity. Rutherford had already discovered that some materials spontaneously emit a kind of radioactive particle, which he called **alpha (α) particles**. They have about seven thousand times the mass of an electron and a positive charge that is twice the magnitude of the electron's negative charge. (Today we know that alpha particles are actually the nuclei of helium atoms.) In 1907, a year after Thomson's proclamation of his "pudding model" of atoms, Rutherford began a long series of experiments. He bombarded a series of targets with alpha particles and observed how they were scattered by viewing the light flashes they produced on a screen coated with zinc sulfide.

Particles with the same electrical charge repel one another, while oppositely charged particles attract one another. If Thomson's "pudding model" of the atom were correct, the positively charged alpha particles would be barely deflected by the diffuse positive "pudding" of the atom. The atom's electrons would also have little effect on the alpha particles, since they are small and widely dispersed. So most of the alpha particles should travel right through the target and be barely deflected.

What Rutherford actually observed, however, was quite different. True, most of the alpha particles passed right through the foil targets with only small amounts of scattering. However, a substantial percentage of them were deflected through some very large angles. Occasionally one would even ricochet right back at the source. Rutherford would later remark that it was "as if you had fired a 15-inch shell at a piece of tissue paper—and it came back and hit you." Thomson's model of the atom was clearly wrong.

Rutherford's experiments suggested instead that an atom consists of a very tiny nucleus where all of the positive charge is located, with electrons occupying the rest of the space in some kind of orbit around the nucleus.

However, this model also had serious problems. The negatively charged electrons and positively charged nucleus must necessarily attract each other. The electrons must somehow continually keep their distance. But it was also known by then that electrons give off energy when they accelerate. An electron in an orbit around an atomic nucleus is subject to a force called centripetal acceleration, so it would radiate energy. The more energy it lost, the closer toward the nucleus it would drop, until it fell into the nucleus itself. That was clearly not happening in real life. Once again, theory was being overwhelmed by hard data. It was up to a Danish physicist to resolve the problem— and in the process to explain the existence of spectral lines.

Niels Bohr was the son of a physiology professor. He received a Ph.D. in physics in Copenhagen in 1911, then spent three years in England. Some of the time was spent with Ernest Rutherford at his laboratory in Manchester. It was there, in 1913, that Bohr formulated a radical new model for the hydrogen atom, cobbling together pieces of classical physics with some new ideas that most people then found nearly unbelievable.

His starting point was Max Planck's quantum effect. Bohr suggested that electrons can occupy "orbits" around the nucleus that had only specific energy levels. The electron orbits are quantized, he said. When an electron absorbs enough energy, *and only then*, it will "jump" to the next higher orbit. When it emits enough energy—and only then—it will jump down to a lower orbit. The jumps are not continuous. They are discrete— instantaneous, in fact. They are "quantum jumps."

This explanation seems at first glance to be absurd. It's as if the space shuttle could fly around the Earth in only, say, three specific orbits—100 kilometers high, 150 kilometers high, and 300 kilometers high. It gets from one orbit to the next not in some continual path upward or downward, but by instantly "jumping" from one to the next when it has enough energy to do so. Objects don't behave this way in the "real world" of space shuttles, rocks, cars, and people. But they do in the microworld of electrons and atoms.

The Bohr theory of electron orbits explains the existence of spectral lines. Electrons emit or absorb energy as they jump from level to level in the atom. (The levels are usually called "shells" by physicists.) Because they do this in discrete amounts or quanta, they produce lines in a spectrum at discrete locations.

Niels Bohr's explanation of spectral lines was only one consequence of his work in creating the theory of quantum mechanics from Max Planck's initial insight. Quantum mechanics would eventually come to be one of the three great pillars of the edifice of contemporary physics. The other two are the special and general theories of relativity, which were created by an absent-minded German who almost flunked out of high school. The three theories do not quite fit together, but all three play a vital role in our understanding of the nature of black holes, including the monstrous one that resides at the center of our Galaxy.

6

FROM RELATIVITY TO THE ISLAND UNIVERSES

If, as most astronomers are now convinced, the center of our Galaxy is home to a huge black hole, it is due to the existence of gravity. Today we know that gravity itself acts in a way that resembles, to a highly accurate degree, a description formulated not by Isaac Newton, but by a former patent clerk named Albert Einstein (1879–1955). This mathematical description is called the theory of general relativity. It's called "general" relativity because, first, Einstein came up with an earlier description of the universe called "special" relativity and, second, it *is* general in the sense of being widespread or comprehensive.

The Bern Patent Clerk and Space-Time

Albert Einstein was born in 1879, the same year as James Clerk Maxwell, Leon Trotsky, Joseph Stalin, and Paul Klee. Henrik Ibsen's play *A Doll's House* was published that year, as was Henry James's novel *Daisy Miller*. L. F. Nilson discovered the element scandium, Australian frozen meat went on sale in London for the first time, and Mary Baker Eddy became the pastor of the Church of Christ, Scientist, in Boston.

Einstein was far from precocious. He dropped out of secondary school in his native Germany, returning only because he couldn't pass the entrance examinations for university without further study. His academic career at the University of Zurich was completely undistinguished. He graduated in 1900. His dull performance at the university kept him from any serious consideration for an academic post. He ended up getting a job as a patent clerk in the Swiss city of Bern. It paid the bills.

It also gave Einstein time to think. And he was a remarkable thinker. Einstein had always had a vivid imagination. He himself once remarked that as a child he had tried to imagine what the world would look like if he were traveling on a beam of light. In the patent office, with its slow pace and undemanding schedule, Einstein began applying his imagination and formidable mental powers to some of physics' most intractable problems. In 1905, at the age of twenty-six, he broke open the world of physics.

Much to their alarm, physicists had uncovered apparent violations of Galileo's fundamental assumption of physics called the **Relativity Principle**. This principle states that the laws of physics should be the same for all observers moving at constant velocities with respect to each other. The laws of mechanical motion formulated by Isaac Newton appeared to agree with this postulate. However, the laws of electricity and magnetism apparently did not. It seemed that the way moving charges affected each other depended on which were moving and which were not. That meant that different theoretical observers of these moving charges would be getting different results—a clear violation of Galileo's relativity principle.

Einstein solved this problem by reaffirming the relativity principle and postulating another fundamental rule: the constancy of the speed of light. As just noted, the relativity principle states that all the laws of nature are the same in all inertial frames of reference. An inertial reference frame is one in which Newton's first law of motion holds true: an object with no net force on it either remains at rest or in motion with a constant velocity. The **constancy of the speed of light** is just that: the speed of light in a vacuum is the same for all observers in inertial reference frames.

Einstein found that these two postulates have some pretty interesting consequences. First, there is no such thing as "absolute motion." There is "relative" motion between two bodies (which Einstein and other physicists call "frames of reference"), but no *absolute* motion. There is no place in the universe where you can stand and say, "This place is solid; it is unique; from here I can measure with certainty the objective motions of plants, stars, people, and that spaceship flying from Earth to the center of the Milky Way Galaxy. I can even objectively compare the motion of objects on that planet, or in that spaceship, and see how they differ." No such place exists.

We've already come across the fact that velocity and speed are measured in units of distance and time. This may seem self-evident, and it is. But this self-evident fact says something extremely important, even fundamental. *Space and time are somehow intimately linked.* They must be. Even in our daily lives, in our commonplace task of driving from home to the supermarket, we experience that intimate connection. To move in space takes time. In 1907 the Russian mathematician Hermann Minkowski (1864–1909) stated his view that space and time should not be seen as separate entities at all, but as different forms of the same thing. Minkowski called this thing **space-time**. The universe, he said, is a *four-dimensional space-time continuum*, with three dimensions of space and one of time. He referred to the sum of all events as "the world" and the path of an individual particle (be it an electron, a planet, a person) in space-time as a **worldline**.

This implies that we can draw a kind of map showing the path or worldline of a particle in space-time. We cannot draw a map in four dimensions any more than a creature that lives in a two-dimensional universe can make a three-dimensional sphere. But we can draw a diagram that measures the three space dimensions in one direction and the fourth time dimension in another direction. Such a drawing is called a **space-time diagram**. Space-time diagrams have become very helpful in understanding what happens in and around black holes, so it's worth taking a closer look at them.

This figure shows the major parts of a space-time diagram. Space is represented by the horizontal line, time by the vertical line. Time proceeds from the bottom (the past) to the top (the future). The two lines at forty-five-degree angles are **light lines**. They represent the worldlines of particles moving at the speed of light.

This sets up an interesting way of interpreting space-time diagrams. Everything made of matter can only travel slower than light. So the worldlines of material particles on a space-time

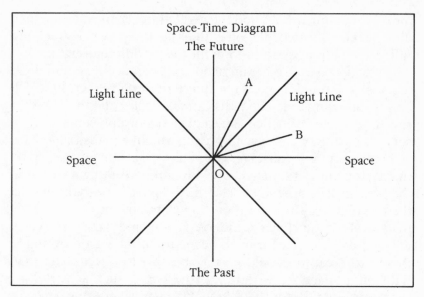

A simple space-time diagram plots all of time and space on two lines. O is the observer. The three dimensions of space are the horizontal line, and time (proceeding from past to future) is the vertical line. Light lines at forty-five-degree angles to the horizontal and vertical lines represent the worldlines of photons moving at the speed of light. OA is the worldline of a particle moving along a timelike path at a speed less than that of light. OB is the worldline of a particle moving along a spacelike path faster than light—something not allowed in our universe. (An unmoving particle would have a completely vertical worldline, since it is unmoving in space, but not in time.)

diagram must be inclined to the vertical time line by less than forty-five degrees. Such worldlines are called **timelike**. Anything that could travel *faster* than light would have a worldline inclined to the vertical time line by more than forty-five degrees and would be a **spacelike** worldline. Special Relativity does not necessarily forbid the existence of particles with spacelike worldlines. However, Einstein's equations suggest that such particles (dubbed **tachyons**) can never travel *slower* than light. So they can never "communicate" with particles traveling on timelike paths—including us.

With all this in mind, we have a space-time diagram like this one. The intersection point for the lines represents the observer (O) at time-now. The horizontal line represents the three space dimensions, and the vertical line stands for the time dimension. The two diagonal lines at forty-five degrees are called the light lines and represent the worldline of light.

This space-time diagram divides all of space-time into three regions. The upper cone is the future of the observer, in which events following timelike paths can take place and the observer can have some influence. The lower cone represents the observer's past, in which events having timelike paths could have taken place and the observer could have had some influence on them. Line OA is a timelike worldline, representing a trip from the present into the future. The third region is everything else, the rest of space-time. It is "off-limits" to anyone and anything that travels on timelike worldlines. That includes the observer at O—and us. In order for us to have any influence on events in this third area, we would have to be able to travel faster than light on a spacelike worldline. Line OB is such a worldline, representing a faster-than-light trip, which nothing (except tachyons, which have yet to be proven to even exist) can take.

Einstein's theory of special relativity states that *all* frames of reference in space-time are the same, provided they are moving at a uniform velocity—neither accelerating or decelerating (constantly increasing or decreasing in velocity) nor rotating. What can be measured is the *relative* motion between two such frames of reference. However, there is no way to determine some "absolute" motion. Two people moving at a constant velocity relative to each other *can never determine which one is moving and which is not.* The laws of physics, including those of motion, are the same for both. There is no absolute motion or velocity to use as a reference point.

For example, imagine you are a passenger in a jumbo jet flying from Seattle to New York. The day is sunny, the air is clear. Five miles below, you can see the Rocky Mountains sliding past. If you wish, you can say that you are stationary and the ground is moving by at nearly seven hundred miles per hour. No experiment you (or anyone else in the plane) carry out will tell you whether that is true or false.

The second postulate also has some remarkable consequences. If the speed of light is the same for all observers in all inertial reference frames, then light is different from any other type of wave we have experienced. That, in fact, turns out to be true. Take sound waves, for instance. Suppose we measure the speed of sound waves (say, from a rock concert by the Grateful Dead) as we are moving toward them. Then we measure the speed of the sound waves as we are moving away from them. We will discover that the velocity of the waves in the first instance is greater than in the second case.

However, this is not the case with electromagnetic waves. Light does not behave in this fashion. It doesn't matter whether you are at rest, moving toward the light beam, or moving away from it. The speed of light is always the same. Period.

There are some other seemingly bizarre consequences of special relativity. The best known are time dilation and shrinking yardsticks. Special relativity says that, as measured by an observer at rest, time runs more slowly in a moving frame of reference. Likewise, an object that is in motion is shorter, as seen by that observer, than an identical object that is at rest relative to the observer. Length contraction is a direct consequence of time dilation, since space and time are two aspects of the same thing (space-time).

Most people have considerable difficulty getting their minds to wrap around these concepts. The reason, of course, is that we never encounter time dilation or length contraction in our daily experience. These bizarre but real effects take place only at relativistic speeds, speeds that are an appreciable fraction of the speed of light. The result is that relativity effects seem to violate common sense. In response to this objection, Einstein is reported to have once said, "Common sense is that layer of prejudice laid down in the mind prior to the age of eighteen."

Mass, Equivalence, and General Relativity

Special relativity deals with measurements of length and time made by observers moving at constant velocities relative to each other. Einstein's theory of general relativity expands to consider accelerated systems and the force of gravity, which were not addressed by special relativity. Einstein published his work on general relativity in 1915.

Like special relativity, this theory also rests on two key elements. For general relativity they are the **principle of equivalence** and the concept of **curved space-time**.

When Isaac Newton formulated his second and third laws of motion, he was talking about two different kinds of masses, or at least two ways of measuring an object's mass. One is called **inertial mass**, and the other is **gravitational mass**. Einstein's principle of equivalence, which he first formulated in 1907, essentially states that both kinds of mass are equivalent. The principle of equivalence is based on Galileo's principle that all bodies, whatever their masses, will accelerate at the same rate when falling under the influence of a gravitational field.

THE INERTIAL MASS OF AN OBJECT

It's easy to figure out the inertial mass of an object. The formula for Newton's second law of motion is

$$F = ma$$

According to this equation, force equals mass times acceleration. Since you know the force applied to the object to get it to move, and the acceleration that then takes place, the object's *mass* must be equal to the force *divided by* the acceleration. In mathematical language, that's

$$m = F/a$$

Simple math isn't *that* scary. Really.

Newton's second law is based on this observation by Galileo. If all bodies falling in Earth's gravity field fall at the same rate, be they balls made of lead or of wood, then the force that is acting on those falling bodies must be proportional to the masses of the objects. The second law therefore states that when a force is acting upon an object, the object will accelerate in the same direction as the force. The force and the object's acceleration are proportional: a two-kilogram mass needs twice as much force as a one-kilogram mass to reach the same acceleration. So the force is equal to some constant of "proportionality" times the acceleration. The proportionality constant is the inertial mass of the object. This kind of mass, inertial mass, is the measure of a body's resistance to a change in its state of motion. If you know the force you are applying on an object, then measure the acceleration that occurs, you can figure out its (inertial) mass.

FIGURING OUT THE FORCE OF GRAVITY

The formula for the force of gravity is

$$F = GmM/r^2$$

It looks scary, but it's really not. F is the gravitational force of attraction. The m and M are the two masses attracting each other, and r is the distance between them, which is squared (the ^{2}). G is a proportionality constant called the **universal gravitational constant**, which determines the strength of the force of gravity. Unless you're a physicist doing serious gravity work, you can ignore it. The point is that you multiply the values of the two masses, then divide your result by the square of the distance between them.

Suppose m is 500, M is 200, and r is 25. Then

$$F = (500 \times 200)/625$$

and F will be 160. Double m to 1,000 or M to 400, and F becomes 320—it also doubles. But if you double r to 50, then

$$F = (500 \times 200)/2500$$

and F becomes 40, one-fourth of 160. If you triple r to 75, F equals 17.77, which is one-ninth of its original value.

Simple.

Gravitational mass, the second measure of mass, is a measure of how much gravitational force the object exerts on another object. Newton's third law of motion deals with this kind of mass. It states, "For every action there is an equal and opposite reaction." (Like Einstein's formula $E = mc^2$, this is one of the few scientific laws that most laypeople recognize and remember.) It follows that if Mass #1 is exerting an attractive force upon mass #2, then mass #2 is also exerting an equal and opposite force of attraction on Mass #1. Not only does the Earth attract the moon, the moon attracts the Earth. Thus, the mutual attractive force must depend on the masses of these two objects.

That's gravitation. It is a force of mutual attraction that is directly proportional to the masses of the attracting objects and inversely proportional to the square of the distance between them. Double the mass of either object, and the gravitational force between them doubles. That's the meaning of "directly proportional." However, the gravitational force is also *inversely* proportional to the distance between the objects. In other words, if the distance between the two objects doubles, then the attractive force *decreases to one-quarter* of its original strength. If the distance triples, the attractive force drops to one-ninth of its original value.

There's no obvious reason why the two kinds of masses, inertial and gravitational, should be the same. However, Galileo's observation of the identical acceleration of falling bodies is possible only if the ratio of gravitational to inertial mass is the same for all objects. Newton carried out a series of experiments to see if the ratio of inertial mass (the second law of motion and the resistance of an object to acceleration) and gravitational mass (the third law of motion, the law of gravitation, and the determination of the accelerating force) were different for different objects. He found no difference.

Nor has anyone else in the centuries since. Experiments that can detect a difference of one part in a trillion have been conducted and found no variation in the ratio. Physicists now conclude that the two kinds of masses are equivalent. This is the principle of equivalence. *Why* these two kinds of mass are equal has been a mystery. The fact that they are is called the *weak* equivalence principle and is what Newton demonstrated.

Needless to say, there is also a *strong* equivalence principle, and it should be no surprise that this is what Einstein demonstrated. One good way to understand it is to do what Einstein himself frequently did: carry out *gedanken* experiments, or thought experiments. The classic thought experiment in this regard is the closed-box experiment.

Suppose you, a kumquat, and a block of lead are locked inside a closed box. You will not be able to tell the difference between the effects of gravity and the effects of acceleration. If the box is resting on the Earth's surface, you will feel your normal weight. If you drop the kumquat and the lead block, they will fall to the floor and will accelerate at the same rate, 9.8 meters per second per second. If you gently toss the kumquat away from you, it will fall to the floor along an arcing path.

Now suppose that you, the kumquat, the lead block, and the closed box are out in deep space. A rocket motor and fuel tank are attached to the box. The rocket is firing, moving the box (plus you and your inanimate companions) at a steady acceleration of 9.8 meters per second, each second. Once again you will feel your normal weight as you stand on the floor of the box. If you drop the kumquat and the lead block, they will again both fall to the floor at the same rate of 9.8 meters per second per second. If you toss the kumquat (or lead block) away from you, it will still fall to the floor along a parabolic path. In fact, all these observations—your weight, the falling acceleration of the objects, the parabolic path the objects take when you throw them—and the observations made when the closed box was safely on the Earth's surface *will be identical.* There is no experiment or observation you can make under these circumstances that can tell you whether these effects are due to gravity or acceleration. They are indistinguishable.

Next, suppose that you, the kumquat, and the lead block are in the closed box, which is now an elevator at the top of a very long shaft. The elevator cable breaks, and the elevator box plunges downward. Everything—closed box, lead block, kumquat, and you—are accelerating downward at the same rate. You are all in free fall, and none of you will experience any weight at all. If you let go of the kumquat and the lead block, they will both float freely in front of you. Question: If this is a *very* long elevator shaft, and everything has been falling since before you were born, how will you be able to tell whether you are really in a free-falling elevator or in a closed box floating freely somewhere in space? Answer: You can't. Once again, there are no experiments or observations you can perform under these circumstances that can make it possible for you to distinguish between a true absence of gravity and a free-falling condition. Newton's three laws of motion will hold true.

The free-falling closed box is a *local* **inertial frame of reference**. An inertial frame of reference is one free from acceleration or rotation, and the free-falling box meets those requirements.

Einstein's *strong* equivalence principle states two things:

1. Gravity and acceleration are indistinguishable in closed boxes.

2. All the laws of nature operate the same way in freely falling boxes as they would in any other inertial frame of reference.

The strong equivalence principle has some remarkable consequences. One has to do with the behavior of light in a gravitational field. If gravity and acceleration are equivalent, then light rays moving through a gravity field will be bent. Also, a beam of light moving *up* through a gravitational field will be redshifted. This form of redshift is called gravitational redshift, and we'll soon encounter it in more detail.

Einstein and Gravity

Isaac Newton's great contribution to understanding gravity was finding a mathematical explanation for this mysterious attractive force between two objects. Newton's inverse-square law describes the behavior of the gravitational force between any two objects. Several important questions, however, were left unanswered: Why should two objects—even when far away from each other—exert a force of attraction on one another? Why should this attractive force follow an inverse-square law? And why is this force apparently transmitted instantaneously across vast distances?

Newton's approach to gravity was pretty much an empirical one: "Here are the experiments; here are the results; here is what the results mean." He was not interested in the *why* of it. "I make no hypotheses" was Newton's answer to those who asked "why" questions.

Einstein was very interested in the "why" of gravity. Soon after he finished his work on the theory of special relativity, Einstein realized that it was not compatible with Newton's inverse-square law of gravity. First of all, Newtonian gravity was a force that traveled instantaneously across all distances. However, a central tenet of special relativity is the light-speed barrier. No material particle or physical disturbance in the universe can travel faster than the speed of light (abbreviated as *c* by physicists). So the Newtonian concept of gravity would have to be modified to conform with special relativity's light-speed limitation.

The second problem had to do with the inertial observer and Newton's concepts of absolute space and absolute time. In Newtonian physics, two events can take place at different locations and times, and the measurements of distance and time intervals between them will be the same for whoever made those measurements. All observers will come up with the same results. This was simple "common sense."

However, special relativity challenged this intuitive view of absolute space and time. Distances and time intervals between two events *do not* have the same values for all inertial observers. Space and time are relative things, relative to the observer in his or her inertial frame of reference. And here is where the second problem arose: *there are no inertial observers of gravitation!* Gravitational forces are everywhere. They permeate the entire cosmos. No observer anywhere is isolated from gravity. The force may be infinitesimally small (say, for example, if you are out in deep space, a million light-years away from the nearest star or galaxy), but it is still there.

An inertial observer, by definition, is one who is either not moving or moving at a constant velocity. However, the nature of gravity is such that it causes an object to be constantly *accelerating* toward the object that is attracting it—even if that object is a star a million light-years distant. So where is the inertial observer of gravitational force? Answer: nowhere. Such a person does not, and cannot, exist.

Einstein solved this apparently insoluble problem in typically radical fashion. He chose to *see* things differently. If gravity is everywhere, all the time, then it can be thought of as somehow fundamentally attached to space-time. Gravity is therefore a basic property of space-time. It is part of its *shape*, part of its geometry. The result of this way of seeing gravity was a new explanation of the geometry of space-time, and of gravity as a geometric characteristic of space and time.

Different Geometries

The geometry that most of us learned in school, and with which most of us are familiar, is the geometry first described by the Greek philosopher Euclid around 300 B.C. Like other branches of mathematics, Euclidean geometry is founded on a few assumptions called postulates or axioms. One of these is that the shortest distance between two points is a straight line. Another is that lines that run parallel to each other never meet. Related to this is a third postulate: if you draw a straight line, then place a point somewhere outside that line, you can draw only one line that passes through that point and is parallel to the first line. Another "obvious" geometrical truth has to do with triangles: the three angles of any triangle will add up to 180 degrees. The truth of these postulates seems obvious. Anyone can take a piece of paper and a pencil and demonstrate it.

However, Euclidean geometry is not the only kind of geometry that is "true." Other geometries exist for which these postulates (and others in Euclidean geometry) are changed. Suppose you are drawing lines on a sphere, not on a flat piece of paper. You will quickly discover that the second postulate just mentioned is *not true*. On a sphere the shortest distance between two points is a straight line that is the arc of a **great circle**. A great circle is the circle drawn onto a spherical surface by a plane that passes through the center of the sphere. It is soon apparent that any two great circles intersect. Two straight lines on a surface are considered parallel if they never intersect, no matter how far out into space they extend. On a spherical surface, however, straight lines *always* meet. Therefore, it is impossible to have two parallel lines on a sphere.

Another Euclidean axiom that is violated on a sphere is the one about triangles. If you draw a triangle on a spherical surface, then add up its three angles, they will total *more than* 180 degrees. These kinds of curved surfaces are said to have **positive curvature**. Other geometries exist that have **negative curvature**. A saddle is a surface with negative curvature. In that kind of geometry the angles of a triangle add up to less than 180 degrees.

The geometry of space and time in special relativity is flat, Euclidean. But when Einstein began working on his theory of gravity, he realized that for gravity to act the way it does, space-time would have to be curved in some non-Euclidean geometry. He used a form of curved geometry called Reimann geometry. In this geometric system, two important Euclidean axioms are still true: a straight line is one whose direction does not change, and the shortest distance between two points is just such a line.

General relativity is a set of equations by which Einstein showed the intimate connection between the geometry of space-time and the distribution of matter in it. Gravity in this concept is *a curvature of space-time*, not a "force" like electromagnetism. When you throw a ball into the air, it falls back to Earth along a curved path. In Newton's conception, the ball moves through flat space-time under the influence of Earth's gravitational force. In Einstein's conception, though, the ball moves through *curved space-time* under the influence of no force at all.

Space-time is like a rubber sheet. Placed on the sheet are marbles of different sizes and densities. The small ones cause small dents in the rubber surface, dents whose curve starts gently a short distance away and doesn't get very steep even at the point where the marble is resting. The marbles that are huge or very dense cause very large dents in the rubber surface. The curvature begins far from the marble and becomes extremely steep at the end. The marbles are matter—dust, comets, asteroids, moons, planets, stars, star clusters, galaxies, galactic superclusters. The dents in the rubber of space-time (which is not a two-dimensional sheet, but a four-dimensional space-time continuum) are gravity. There is no place in the universe where space-time is not curved. Even if all the cosmos contained only one tiny speck of matter, that dust mote would create a gravitational warp in all of space-time.

The curved space-time of general relativity has some interesting consequences for the nature of things we take for granted in our Euclidean-perceived reality. For example, a straight line is not really straight at all, at least not as we imagine it. "Straight lines" cannot exist in a cosmos where space itself is curved. A straight line is always moving in the same direction. It's just that space itself is curved. So "straight lines" are always curved. (But curved lines are not straight; they're curved, too, but more curved than straight lines.)

Then there is the matter about the shortest distance between two points. It is still a straight line—but not as we usually define "straight." One way to picture this is to plot the route between two cities on a globe. Now plot the route between the same two cities on a flat map drawn to the same scale. The route is different. The Earth is a curved sphere, and no flat map—not even the best made by the National Geographic Society—can perfectly reproduce the geometry of a sphere. A straight line on a globe must be depicted as a curved line on a flat map. It doesn't *look* like the shortest distance between the two points— but it really is. The same is true of lines in curved space-time. A picture of an object moving in space-time—a space-time diagram—plots movement in space along one axis and movement in time along the other. The result is a worldline spiraling upward from time-present to time-future.

A worldline that is a depiction of straight-line movement in curved space-time is called a **geodesic**. No other force is acting upon the object traveling along the geodesic. That includes gravity because, as defined by Einstein in the theory of general relativity, gravity is not a force but a characteristic of the geometry of curved space-time.

Solving the Mercury Puzzle

A theory is only as good as the accuracy of its predictions. General relativity soon proved how accurate it was by explaining the mystery of Mercury's orbital precession. Mercury is the planet closest to the sun. Like the orbits of the other planets, the axis of Mercury's orbit itself rotates around the sun in the phenomenon called precession. This means that Mercury's perihelion (the point in its orbit closest to the sun) and aphelion (the point farthest from the sun) do not stay in the same place in space relative to the sun. They move slowly but steadily around the sun. The orbital precession of the other planets in the solar system can be easily explained by the equations of Newtonian gravity. However, Mercury's orbital precession is much larger than the Newtonian equations predict. This was a conundrum for astronomy, a monkey wrench in the steady clockwork mechanism of the Newtonian view of the universe.

Newton "explained" Mercury's orbital precession by invoking the Deity. The mystery was actually unraveled by general relativity. Einstein's equations, as first solved by German scientist Karl Schwarzschild in 1916, predicted that Mercury's perihelion would move around the sun at an angular rate of forty-three minutes of arc per century. That is almost exactly the actual rate of precession for Mercury's orbit.

The solution to the mystery of Mercury's perihelion was impressive, but for many astronomers and physicists not entirely convincing. The difference between the Newtonian solution and the relativistic solution was small. Some larger and more spectacular experiment was needed. Such an experiment took place on May 29, 1919, when the sun went dark.

Gravitational Bending

By 1919 Max Planck's quantum explanation of the nature of photons of light had become widely accepted by the physics community. In his theory of special relativity, Einstein had shown that energy and mass are different forms of the same thing: $E = mc^2$. Photons may be particles that have no mass, but they do have energy, and that energy is the equivalent of mass. (Simply divide the amount of the energy by the square of the speed of light, and you have the mass.) Therefore, a beam of light passing very close to the surface of a massive body should be bent by the object's gravitational pull.

Another way to look at it is with the geometric analogy. The presence of a massive body in space and time will bend space-time. A beam of light traveling very close to that massive body will naturally follow a curved path past it. Such a thing was not considered possible under Newton's theory of gravity; light was supposed to be immune to gravitational effects. This clear distinction between the two theories inspired British astronomer Sir Arthur Eddington to come up with a way to prove or disprove the validity of general relativity.

Eddington was the Carl Sagan of his time, a well-known astronomer who was also a good writer and excellent popularizer of science and astronomy. Eddington knew that if the gravity field of the sun bent light rays passing near its surface, then the images of stars that would lie near the sun's limb—its visible edge—would appear shifted farther away from the limb. A star that was actually barely hidden by the sun's limb would appear just above the limb. The light rays from this star would be bent around the sun by its gravitational pull. Thus, the star would appear to be just above the sun's visible limb. Of course, there is no way to see stars near the sun's limb . . . except during a total solar eclipse.

Eddington realized this could be a perfect "natural experiment" to test general relativity. Such an opportunity would present itself fairly soon. A total solar eclipse would take place on May 29, 1919, off the coast of South America. Eddington persuaded British Astronomer Royal Sir Frank Dyson to lend support to two expeditions that would observe the eclipse at Sobral, Brazil, and on the island of Principe in the Gulf of Guinea. The observations made that day during the solar eclipse supported the predictions of general relativity regarding the gravitational bending of starlight.

For many years the 1919 solar eclipse was presented as *the* proof of general relativity. It certainly appeared so. In fact, it was considered an event of such great importance that the story made the front page of the *New York Times*. Many years later, however, a fly appeared in the relativistic ointment. It turns out that more standard effects of light refraction take place near the sun's surface. These refraction effects cause a bending of visible light rays very close to that predicted by general relativity. So, what was the *real* cause of the starlight bending seen in 1919— gravitational effects à la general relativity, or refraction?

The answer finally came in 1975 and 1976. The refraction effects, while strong for *visible* light, are nearly negligible for radio waves and microwaves. Scientists therefore measured the bending of light at those wavelengths from a distant quasar as it was blocked by the sun passing in front of it (a common astronomical event called an **occultation**). The sun itself produces relatively few radio waves and microwaves, so the experiments did not have to take place during a solar eclipse. The results of these observations clearly proved that the gravity field of a massive object *does* bend light rays. General relativity is correct.

The gravitational bending of light seems strange enough, but it is not the only consequence of general relativity and the concept of curved space-time. Another is **gravitational redshift**. Suppose that a beam of light with a particular wavelength is leaving the surface of a very massive star. You are on the surface of a small planet observing the light as it reaches you and measuring its wavelength. The equations of general relativity predict that the gravitational field of the massive star will cause an *increase in the time span between each crest and trough* of the light wave. That in turn means that you will see an increase in the wavelength of the light. The shorter the wavelength of light, the bluer it is; the longer the wavelength, the redder it is. (In fact, it is perfectly correct to say that radio waves are "redder" than red light and that x-rays are "bluer" than violet light. They are. If you are a honeybee, for example, you can actually see infrared light as a visible form of light. It would be "redder" in color than red light.) This shift in wavelength toward the redder part of the electromagnetic spectrum is called a gravitational redshift. Such redshifts have been seen in the spectra of extremely massive stars called white dwarfs.

But black holes are even "denser" than white dwarfs. We will soon see how gravitational redshift plays a role in understanding these bizarre holes in space-time and the laws of nature.

The Great Debate and the Island Universes

Galileo's crude refracting telescope in 1610 was not much more than a good spyglass—but it was enough to reveal astronomical phenomena that led to the final overthrow of the Ptolemaic worldview. By the 1780s William Herschel was building reflecting telescopes of unparalleled size and resolving power. For example, he eventually built and used a telescope that had a mirror 1.2 meters (4 feet) in diameter. With telescopes of this (but usually somewhat smaller) size, astronomers continued to discover new objects in the universe.

In particular, it became clear that the skies were home to various patches of fuzzy-looking luminous objects called **nebulas** (from the Greek *nephos*, meaning "cloud"). More than a hundred of these were listed in the eighteenth century by Charles Messier in what is now called the Messier catalog. Some of these nebulas appeared to be amorphous-looking blobs of glowing gas. They were eventually identified as just that— clouds of gas and dust that were clearly inside the Galaxy. Others turned out to be groups of stars, also within the confines of the Galaxy. Still other nebulas had a spiral shape and were called spiral nebulas.

For well over a century, astronomers debated the question of what spiral nebulas actually are. Some believed that, like the other kinds of nebulas, these are clouds of gas within the confines of the Galaxy. Others, however, argued that the universe is much larger than our own Galaxy and that spiral nebulas are themselves "island universes" of stars like our own Milky Way. The question, in other words, was really two questions: What is the nature of the spiral nebulas, and how big is the universe?

The controversy came to a head in 1920. That same year Arthur Stanley Eddington wrote his book *Space, Time, and Gravitation*; Warren G. Harding was elected president of the United States; Ralph Vaughan Williams completed the final version of his *London Symphony*; and Harvard beat the University of Oregon 7–6 in the Rose Bowl. On April 26 of that year, two well-known astronomers, Harlow Shapley and Heber D. Curtis, addressed the controversy in a public meeting at the National Academy of Sciences in Washington, D.C. Most popular astronomy books and texts today speak of this encounter as a "Great Debate" between the two leading proponents of the opposing views. This is a somewhat romanticized version of events. Still, the debate (which was not a debate) served to crystallize the two competing astronomical worldviews of that time. In the end, both Shapley and Curtis turned out to be wrong—and right.

Up through the first decade of the twentieth century, the prevailing view was pretty much that of Herschel some one hundred years earlier: the Milky Way Galaxy contains the sum total of the stars, gas, and dust in the universe. Whatever lies beyond it, if anything, is empty darkness. The spiral nebulas are therefore small objects located in the Galaxy itself.

By 1920, however, that view had largely changed. The Galaxy was by then considered to be a relatively small collection of stars and nebulas, with the sun located somewhere near the center. One version of this view, the so-called "Kapteyn universe" (named for the Dutch astronomer Kapteyn, who championed it), held that the Galaxy has an ellipsoidal structure. The nature of and distances to the spiral nebulas were unknown. The more prevailing view held that the Galaxy is spiral in shape and that spiral nebulas are distant island galaxies like our own.

In 1918 the young and ambitious Harlow Shapley (who worked at the Mount Wilson Observatory in California) proposed another, more radical theory. He proposed that the Galaxy is quite large in extent, at least three hundred thousand light-years in diameter, and that spiral nebulas are little more than nearby nebulous objects. This was obviously in stark contrast to the then-prevailing view. Heber Curtis, then an astronomer at Lick Observatory, was a tenacious defender of the so-called "island universe" theory.

In late 1919 George Ellery Hale (the director of the Mount Wilson Observatory) suggested to the National Academy of Sciences that one of its upcoming lectures established in the memory of his father be devoted to either the island universe controversy or to relativity. The academy settled on the former and eventually on Shapley and Curtis as the two opposing speakers.

The actual debate was not only not a debate, the two speakers gave entirely different kinds of speeches. Shapley gave a talk that was elementary and mostly popular in tone. Its main thrust was not a direct attack on the concept of separate spiral galaxies but a defense of his "giant Galaxy" theory. By contrast, Curtis had come prepared to give a fairly technical talk illustrated with slides and charts, defending the island universe theory and attacking Shapley's theory. Shapley went first, and Curtis was quite surprised by what he saw and heard. Nevertheless, he went ahead and gave his talk as he had originally prepared it. Neither principal speaker had the right to reply to the other, so it wasn't much of a debate at all. What's more, Shapley attempted to sway matters a bit in his direction by "planting" his friend and mentor Henry Russell in the audience to ask the first question. Never let it be said that astronomers aren't willing to manipulate events to their own advantage! It should be pointed out, though, that Shapley and Curtis were good friends both before and after the "Great Debate."

The event turned out to be quite entertaining for both the sponsors and the attendees. The two were invited to repeat their performance for the 1921 meeting of the Astronomical Society of the Pacific. They gracefully declined. Once was enough for these two competitive but friendly colleagues.

In the end there were two great ironies to the "Great Debate" over island universes and the size of the Galaxy. The first was that the debate solved nothing. In 1920 there was simply not enough hard information to settle the matter one way or another. The resolution came four years later with a discovery by the great astronomer Edwin Hubble. He found a particular type of star in M31, the Andromeda Spiral Nebula, that allowed him to directly establish a distance from the sun to that nebula. That distance was, by his estimation, some 490,000 light-years. That immediately settled the controversy. The Andromeda Nebula was far beyond the boundaries of the Milky Way Galaxy, even by Shapley's inflated standards. Today we know it as the Andromeda Galaxy. It is a spiral galaxy much like our own. We also know now that Hubble's first distance estimates were quite low. M31 is actually about 2.2 million light-years away.

The second irony? It is truly that. It turns out that our Galaxy *is in fact larger than even Shapley thought*!

And not even Harlow Shapley—a brilliant astronomer who was also an excellent writer of popular science books, the Isaac Asimov of his day—suspected that the Galaxy might hold at its center an invisible object as massive as a hundred million suns.

7

BLACK HOLES IN TIME AND SPACE

The phrase *black hole* was first coined in 1968 by physicist John Wheeler of the University of Texas, one of the world's two or three acknowledged experts on these bizarre objects. But the first mention of a "black hole" took place 185 years earlier. That honor goes to an Englishman, an Anglican priest and amateur astronomer named John Michell.

From Michell to Schwarzschild

The Rev. Michell was not a "professional astronomer"—in the late seventeenth century few such people existed. Astronomy as a full-time profession was only beginning to develop, mainly through the brilliant work of people such as Pierre-Simon Laplace in France and William Herschel in England. However, Michell was far from an unknown. Some years earlier, for example, he had argued persuasively that double stars are too common to be caused by chance alignments in the sky of unconnected stars. Some of them must be physically bound by gravity. Herschel confirmed this. So his suggestion of the possible existence of invisible stars was not dismissed by his contemporaries.

Michell first mentioned the idea in 1783, in a letter he wrote to Henry Cavendish, the physicist and chemist who discovered the element hydrogen. Cavendish in turn reported it to a meeting of the Royal Society of London on November 27 of that year. Michell's letter was published as an article the following year, in volume 74 of the society's *Philosophical Transactions*.

Michell was not looking for black holes or trying to prove

131

that such an object might exist. He was trying to figure out some different ways of determining the distance and brightness of stars. The full title of his letter to Cavendish was: "On the Means of discovering the Distance, Magnitude, &tc, of the Fixed Stars, in consequence of the Diminution of the Velocity of their Light, in case such a Diminution should be found to take place in any of them, and such other Data should be procured from Observations, as would be further necessary for that Purpose."

What Michell suggested was that the properties of some stars might be found by measuring the effect that gravity might have on the velocity of the light a star emits. In the process of developing this theory, Michell made some interesting mathematical observations. He calculated that the escape velocity from the sun is about five hundred times less than the speed of light. Therefore, if the sun had its same density but a diameter five hundred times as great, its gravitational field would trap all the light it emits. Michell's exact words were:

> Hence, according to article 10, if the femi-diameter of a fphere of the fame denfity with the fun were to exceed that of the fun in proportion of 500 to 1, a body falling from an infinite height towards it, would have acquired at its furface a greater velocity than that of light, and confequently, fuppofing light to be attracted by the fame force in proportion to its vis inertiae, with other bodies, all light emitted from fuch a body would be made to return towards it, by its own proper gravity.

(The use of *f* for *s* in many words was quite common in 1784.)

Such an object could never be directly observed, Michell noted. It could be detected only by its gravitational effects on other nearby objects. William Herschel was so taken with the concept that he futilely tried to detect the existence of "retarted starlight." He did not see any, and unfortunately for Michell's place in astronomical history books, his 1784 paper was soon forgotten.

Thirteen years later the concept cropped up again. Pierre-Simon Marquis de Laplace (1749–1827), the famous French astronomer and mathematician, spoke of such an object in his book *Exposition du système du monde*, published in 1796. Laplace postulated the existence of a star so massive that its gravitational pull produced an escape velocity greater than the velocity of light. If such an object existed, he wrote, its light would never escape from it. Thus, it would be an invisible star.

However, like Michell's comment on invisible stars, Laplace's writings on the subject were also mostly ignored. Thus, the early history of the idea of a stellar black hole is one of rediscovery followed by resistance, rejection, and forgetfulness.

The black hole idea next resurfaced in December 1915, 119 years after Laplace's comments. Karl Schwarzschild was born in 1874 and was one of Germany's most brilliant astronomers. He was also quite patriotic. When World War I broke out, he volunteered to serve in the army, even though he was already forty years old. He served in Belgium and France, and in 1915 was transferred to the Russian Front. It was there that he contracted pemphigus, a rare and often fatal disease characterized by the appearance of large and painful fluid-filled blisters on the skin. Throughout all this, Schwarzschild continued doing science. He had just read the four papers published by Albert Einstein a few months earlier, which set forth Einstein's theory of general relativity. So what he did was work out the first complete solutions to those equations.

This would not have been an easy task under even the most favorable conditions. There are sixteen such equations, each extremely difficult to understand and carry out, and the solutions for each can be interpreted in several different and subtle ways. Einstein himself was content to settle for only approximate solutions to his own equations. Schwarzschild, however, pressed forward, despite the chaos of war and the agony of his now-terminal disease. The result of his work was a paper entitled "On the Field of Gravity of a Point Mass in the Theory of Einstein."

Schwarzschild sent his solution to Einstein, who was very happy with it. Wrote Einstein to Schwarzschild: "I have read your paper with great interest. I had not expected that the exact solution would be formulated so simply." Einstein recommended that the Royal Prussian Academy of Sciences publish it, and (not surprisingly) his recommendation was followed. The paper was published in the 1916 edition of the *Journal of the Royal Prussian Academy of Sciences.* Schwarzschild never saw it in print. Two months after finishing the paper, he was sent home to Germany. Two months after that, he was dead.

The paper was only a few pages long, mostly mathematics; there is no evidence that Schwarzschild himself thought he had stumbled upon anything remarkable. He simply set forth his method of solving Einstein's general relativity equations using the example of "a point mass." The paper aroused no stir of controversy or excitement when it was published.

This was not because Schwarzschild's work was unimportant. It was revolutionary. His solution to the general relativity equations describes a gravitational field *outside* of a point mass. The point mass may represent a ball of lead, the Earth, the sun, a superdense star, a spherical collection of stars—anything spherical. However, the general relativity equations were (and are) so difficult to work with and understand that in 1916 almost no one in the physics community appreciated the implications of what Schwarzschild had done.

Einstein did. He realized that one consequence of Schwarzschild's solution was the existence of a point with infinite density. Einstein was a great admirer of Schwarzschild. He even wrote an obituary for Schwarzschild, which appeared in the same issue of the journal that carried Schwarzschild's paper. But Schwarzschild's solution was too strange for even Einstein to accept. He later wrote a paper attempting to show that Schwarzschild's solution to his equations could not have any connection to the real world. What took Schwarzschild only a few months to do, took physicists more than sixty years to understand. For what Schwarzschild described in his small paper was a black hole.

A Black Hole Thought Experiment

Most astronomers call a black hole "black" because light cannot escape from it. (Most, but not all. In particular, Russian scientists call black holes "frozen stars." This difference in nomenclature is a fascinating example of how language reflects differences in cultures. Russians already use the term *black hole*, but not in any technical sense. In their language it is an obscene slang term for the human anus. Thus, they use the phrase *frozen star* to refer to a black hole.) A black hole's escape velocity is greater than the speed of light. **Escape velocity** is the velocity that an object (a rock, a spaceship, a planet) must have to escape the gravitational pull of another, more massive object (a planet, a star, a galaxy). The escape velocity for Earth is about 7 miles or 11.2 kilometers per second. Throw a pebble into the air, fire a bullet from a gun, and they will fall back to Earth. Fling a rocket into space at a velocity greater than 11.2 kilometers per second, and it will permanently escape the gravitational pull of the Earth.

The sun's gravitational force is much greater than that of Earth. If a person weighing 68 kilograms (about 150 pounds) on Earth were to stand on the surface of the sun (assuming the sun's visible surface were a physical surface, which it is not), that person would weigh nearly 2 tons. He or she would also have a much harder time leaving the sun, since its escape velocity is about 617 kilometers per second.

One way to understand what a black hole is and why it is "black" is to carry out a thought experiment. Suppose the sun were to begin collapsing. It has a radius of 696,000 kilometers. When it reached half its present size, about 348,000 kilometers, escape velocity at its visible surface would be about 880 kilometers per second. At one-tenth its present radius (69,000 kilometers), the sun's escape velocity would climb to more than 1,600 kilometers per second. When the sun reached the radius of the Earth, about 6,400 kilometers, its escape velocity would be more than 6,000 kilometers per second. Finally, when the sun collapsed to a radius of about 3 kilometers, its escape velocity would be nearly 300,000 kilometers per second. That's the speed of light. Not even the sun's own light would be able to escape from it.

Any object that has collapsed to the point where its escape velocity is equal to the speed of light cannot be seen at any electromagnetic wavelength—visible light, ultraviolet radiation, radio waves, x-rays, nothing. It is "black." Nor can anything that falls onto such an object ever escape, for nothing can travel faster than light. That object is indeed a "black hole."

Black Holes and Einsteinian Gravity

This is how we might understand the nature of a black hole using Isaac Newton's theory of gravity. Newton's gravitational theory is now known to be a limited case within the larger and more comprehensive theory of gravity developed by Einstein in his general theory of relativity. In truth, the black hole of Newtonian gravity is not the same as the black hole of general relativity. This is because the escape velocity explanation is not really accurate. If we launch a rocket that does not have enough velocity to escape Earth's gravity, it will still travel quite a distance from the Earth before our planet's gravitational pull brings it around in a parabolic curve and it falls back to the ground. In the same way, Newtonian gravitational theory would allow particles of light to travel a certain distance away from a black hole before its gravitation pulled them back. In the real world of black holes ruled by general relativity, however, this cannot always happen. Light emitted from the "surface" of a black hole cannot move outward at all! In either case, however, a black hole is black because no matter or light or any other form of electromagnetic radiation can ever escape from it.

Even though no light, and therefore no energy, can escape from a black hole, these bizarre objects can still be powerful sources of energy in the universe. This is because of their immense gravitational fields. Suppose a black hole with the mass of ten suns is surrounded by a cloud of interstellar gas and dust. Some of the matter will be far enough from the black hole that it will either escape its gravitational pull or merely go into orbit around it. However, the black hole's gravitational field will undoubtedly capture some of the matter and begin pulling it inexorably in. As it falls toward the black hole, the matter will build up enormous amounts of kinetic energy. As the clouds of dust and matter collide, some of that energy will be turned into heat energy. Still more will be released in the form of electromagnetic radiation such as visible light, x-rays, and gamma rays.

The matter accumulates in a disklike structure called an **accretion disk**. The accretion disk lies in the equatorial plane of the black hole, much as Saturn's rings lie in the plane of its equator. Accretion disks are much larger than the black hole itself. A black hole with the mass of several suns, for example, may have a radius of only thirty kilometers, but its accretion disk could be as big as our entire solar system. Various physical phenomena also make it possible for jets of matter and energy to spew off the accretion disk in directions perpendicular to it, along the axis of its rotation. These jets thus appear to be coming out of the black hole's north and south polar regions.

Black holes themselves may be invisible, but they leave their calling card. We cannot see the hole itself, but we can see its accretion disk and can detect the enormous amounts of energy that are being released as the matter in the accretion disk spirals into the black hole. That means we may be able to detect their presence in the universe, and even in our own Galaxy.

Black Hole Characteristics

For most spherical objects, there is nothing unusual about the Schwarzschild solutions to the general relativity equations. Their gravitational fields are quite normal. That's not the case with black holes, though, and any spherical body can be turned into a black hole if we compress it enough. As the sphere becomes smaller and smaller, its radius and surface area decrease, the matter of which it is made is squeezed into a smaller and smaller volume, and its density increases. The intensity of gravity at its surface gets greater and greater. Eventually it is so intense that not even light can escape from the surface. The sphere becomes a black hole.

In the case of the sun, that occurs when its radius is about three kilometers. This is the **Schwarzschild radius** of the sun. An object's Schwarzschild radius is that at which the object's surface escape velocity is equal to the speed of light. The sun's density at this moment would be about 1×10^{16} grams per cubic centimeter. (The density at the center of the sun is only about 160 grams per cubic centimeter.)

The density of any solid object is the ratio of its mass to its

SOME EXAMPLES OF SCHWARZSCHILD RADII
AND BLACK HOLE DENSITIES

Object	Mass	Density at the Moment the Object Reaches Its Schwarzschild Radius*	Schwarzschild Radius
Asteroid	10^{21} g	10^{41} g/cm^3	10^{-12} cm
The Earth	6×10^{27} g	10^{27} g/cm^3	1 cm
The sun	2×10^{33} g (1 solar mass)	10^{16} g/cm^3	3 km
Supermassive star	10 solar masses	10^{14} g/cm^3	30 km
Supermassive object at Galactic center	10^8 solar masses	1 g/cm^3	3×10^8 km
The Galaxy	10^{12} solar masses	10^{-6} g/cm^3	0.3 light-years

*Approximate values, rounded off to the nearest power of 10.

volume. The volume of a sphere depends on the cube of its radius. In the case of a black hole, the radius is the Schwarz-schild radius, and it depends entirely on the black hole's mass. So the greater the mass of the object that becomes a black hole, the smaller will be the black hole's density. The Schwarzschild radius of Earth is about one centimeter, about one-third of an inch. The density of an Earth-mass black hole would be about 1 $\times$ 10^{27} grams per cubic centimeter. A star of ten solar masses that collapsed into a black hole would have a Schwarzschild radius of about thirty kilometers and a density of about 1 $\times$ 10^{14} grams per cubic centimeter. A black hole with about one million to a hundred million solar masses, of the kind that might be present at the center of our Galaxy, would have a Schwarzschild radius of about 300 million kilometers and a density of water. If the entire Galaxy were compressed into a black hole, its Schwarzschild radius would be about three-hundredths of a light-year, some 283 billion kilometers or about forty-eight times the radius of the solar system. Its density would be very

BLACK HOLES IN TIME AND SPACE   139

low, only one-millionth of a gram per cubic centimeter—about the density of air.

The densities given for these black holes, of course, are not constant. They are the densities of the object *at the moment it collapses past its Schwarzschild radius.* No force in the universe can stop the collapse, so the density of the object eventually climbs to infinity.

The Schwarzschild radius is more commonly called the **event horizon**. Here on Earth, the horizon is the limit of our range of vision caused by the curvature of the Earth's surface. As long as we stand in one place, we cannot see what is happening beyond the horizon. In the case of black holes, the curvature is of space-time itself. It is caused by the intensely concentrated mass of an object. The curvature is so great, in fact, that an outside observer can never see what is happening beyond it. Whatever is within this kind of horizon is forever hidden from external view. It doesn't matter where we stand, or how close we get to the event horizon, or how powerful our telescopes might be, or what kind of gee-whiz-Star-Trek sensor we use. The event horizon is as close to an absolute boundary as anything can be in the cosmos.

This means that, unlike stars or galaxies, black holes are very simple objects. Stars, for example, can be characterized by their color, gravitational strength, luminosity or intrinsic brightness, magnetic field strength, mass, size, spin rate, and temperature. But properties like color, luminosity, and temperature are all lost when a supermassive star becomes a black hole. The only properties a black hole has, in fact, are its mass (which scientists symbolize by the letter M), its spin (J), and its electrical charge (Q). This was humorously expressed by John Wheeler when he said: "A black hole has no hair."

A black hole has no other features than these three. A black hole made from a million tons of hydrogen and one made from a million tons of canceled checks will be identical, assuming they also have the same spin or charge. The mass of a star is not lost when it becomes a black hole. It remains, along with the gravitational force it produces. Light cannot get past a black hole's event horizon, but its gravitational field can. So can any electromagnetic field caused by a net electrical charge, which the star might have possessed before its collapse. If a star had an electrical charge at the moment of its collapse, the resulting black hole will, too. A black hole can also have spin. If the star was spinning when it imploded (and nearly every astronomical object has some spin or rotation), so will the black hole.

As we've seen already, it is also possible to speak of a black hole having a radius or diameter. However, this is really not the radius of a physical object, but the radius that existed at the moment its surface escape velocity reached light speed. When astronomers or physicists speak of "the radius of a black hole," they are actually speaking of the radius of the object at the moment its surface passed through the event horizon. *The object itself* continues to become smaller and smaller. Its radius continues to decrease until it apparently reaches zero length and the curvature of the object's surface supposedly becomes infinite. The result is what astronomers and physicists call a *singularity*, a zero-dimensional point of infinite density where all the laws of the universe cease to function.

This whole idea of something possessing zero length, infinite curvature, and infinite density makes scientists somewhat uncomfortable. The reason is that at this point the object is no longer subject to the laws of nature. Those laws cease to apply to a singularity. Anything—anything at all—could be happening in a singularity. A scientist cannot make predictions of what might occur next in such a place or what might emerge from a singularity. However, there is (as far as we know) simply no way to know what *is* going on in a singularity.

The event horizon of a black hole is like the veil that Moses wore over his face after he had gazed upon the countenance of God, as recounted in the Biblical book of Exodus. Moses's face shown with such intense light that no human could gaze upon his unveiled face and live. In somewhat the same way, the black hole's event horizon is a veil across the face of the singularity. Since nothing can escape past the event horizon, there is no way in which we can ever gather any information about what is happening at or in the singularity.

This is called the **cosmic censorship principle**, and it

was first formulated by British physicist Roger Penrose. Penrose convincingly showed that if an object undergoes a spherically symmetrical gravitational collapse into a black hole, it will *always* form an event horizon. The event horizon hides the singularity at the center of a black hole from any observation by outsiders. A singularity that could be directly observed is called a **naked singularity**. A naked singularity, should one ever exist, would make it impossible for scientists to ever confidently predict what might happen in the universe. *Anything* could come barreling out of a naked singularity to throw awry any scientific prediction. Fortunately for scientists, the event horizon "censors" a naked singularity, as it were.

The cosmic censorship principle holds that the universe is constructed in such a way that event horizons always form around black holes and decently hide singularities from the law-abiding cosmos. However, no one knows for sure whether this is really always the case. As we shall see, it might be possible for some kinds of black holes to lose their event horizons.

Kinds of Black Holes

Four basic kinds of black holes are recognized by scientists and are named for the scientists who first worked out the mathematical descriptions of them. They are the **Schwarzschild black hole**, the **Reissner-Nordström black hole**, the **Kerr black hole**, and the **Kerr-Newman black hole**. Each type of black hole has a different set of properties.

There are several ways to explore the internal structure of these four kinds of black holes and to examine the area surrounding them. One way is with a beam of light. We can take a flashlight, for example, direct its beam around the region near a black hole, and see how the curved space-time bends its light. Another way to probe the region of space-time around a black hole is with particles of matter. We can put pieces of matter— say, a spaceship or two astronauts—in the vicinity of a black hole, then see how their motion is affected. Each kind of black hole has a slightly different "internal structure" and affects particles and light rays in different fashions.

The Schwarzschild Black Hole

The Schwarzschild black hole is the simplest form of black hole. As complicated as are the mathematics for understanding black holes, this kind remains the most extensively studied. It is named for Karl Schwarzschild, the German physicist who first solved Einstein's general relativity equations in 1916. It is a nonspinning object with no electrical charge. Einstein himself was quite disturbed by this consequence of Schwarzschild's solutions. He spent several years trying to prove that black holes could not exist. He did not succeed. Unlike other kinds of black holes, a Schwarzschild black hole has one event horizon. It also has one photon sphere. To uncover what a photon sphere is and to examine these characteristics of a Schwarzschild black hole, we will probe it with particles and light beams.

First, the particles: Suppose two space-suited astronauts, Chris and Barbara, step out of a spaceship orbiting the Schwarzschild black hole. Each wears a wristwatch, and the watches are synchronized. Unfortunately, Barbara accidentally triggers the rocket jets on her backpack. She begins falling out of orbit into a nonstop plunge to the black hole.

As Chris watches, Barbara accelerates toward the Schwarzschild black hole, just as she would if she were falling toward the Earth. But then Chris will observe that Barbara appears to be slowing down. This is because Barbara's time, as seen from Chris's frame of reference, is moving more slowly. As Chris continues to watch, Barbara appears to fall more and more slowly as her time moves more and more slowly. It seems logical that (from Chris's point of view) when Barbara reaches the black hole's event horizon, she will come to a complete stop because of the time-dilation effect. It will be as if she is hovering just above the black hole's event horizon forever.

However, another effect of relativity also comes into play at this point—gravitational redshift. The wavelengths of radiation coming from Barbara (which are how Chris sees her, of course) will be more and more redshifted as she falls closer and closer to the event horizon. The frequency of the radiation declines ever more swiftly. The time between the next to the last wavelength and the final one will become infinitely long. So, from

Chris's point of view, Barbara will indeed vanish right after falling past the event horizon.

Events take place rather differently from Barbara's frame of reference. She will not experience any slowing down at all. In fact, as far as Barbara can tell, she will fall right into the Schwarzschild black hole rather quickly. Time runs normally. When she looks back at Chris, though, he will appear to be moving faster and faster. So will everything else in the universe. As she gets closer and closer to the event horizon, the universe and everything in it (including Chris) will rapidly age. Chris will die, and so will the entire cosmos.

These bizarre temporal effects are caused by the extreme curvature of space-time by the black hole. Not only is space warped, so is time. The time experienced by Barbara, the one falling into the black hole, is called **proper time**. The time experienced by Chris, who is observing Barbara falling into the black hole, is called **coordinate time**.

The time warping is not the only thing that happens to Barbara. This is a stellar black hole she is falling into. Space-time in the region of the event horizon is highly distorted, and the gravitational force is so strong that it causes powerful tidal effects. As she falls feet first into the Schwarzschild black hole, her feet experience a stronger gravitational pull than her head. The difference causes her body to quickly get stretched thinner than a piece of string. Obviously, Barbara will not survive her plunge into this black hole.

Were the black hole much more massive than just one star, however, things would be different. Suppose she were falling into a supermassive black hole, one with the mass of a million stars or so. It would still have an event horizon, a region where light is no longer able to escape from the object. However, this region is much more distant from the black hole's singularity. Space-time is less distorted, gravity is considerably less powerful, and tidal forces less extreme. In fact, the tidal forces at the event horizon of a hundred-million-solar-mass black hole are less than the tidal effects we experience every day at the Earth's surface from our planet's own gravity. Barbara could easily plunge past the event horizon of such a supermassive black hole without even noticing any tidal effects.

What would Barbara see as she fell past the event horizon of a black hole? Some very strange things. As Barbara looks back at the event horizon just after she passes through, it will appear to be receding from her at the speed of light. Meanwhile, she will continue her plunge toward the singularity at the center, and there she might be crushed into infinite density.

—Or maybe not. No one can say for sure, since no one knows exactly what lies at the center of a black hole. Indeed, there is reason to believe that a point of infinite density does not—and perhaps cannot—exist. As the matter of the collapsing star gets closer to the region of infinite density, its size becomes so tiny that it reaches the point where the laws of quantum physics become dominant. In regions of space-time smaller than a quark (the "fundamental particle" of which protons and neutrons are made), nothing is certain. Space-time itself resembles the froth on a head of beer. It bubbles and foams, as energy and subatomic particles pop in and out of existence. Einstein's theory of general relativity is not complete enough to predict what takes place at this level of existence. What is needed is a combination of relativity and quantum mechanics, and that theory of physics does not yet exist.

Thus, there is as yet no way to predict with certainty what lies at the center of a black hole. That's as good a reason as any to continue calling whatever it is a singularity. It is at this time singularly unknowable. For the sake of simplicity, most astronomers and physicists continue to refer to it as a "point of infinite density," even though it may well not be so. For the same reason, they assume that singularities are not good places upon which to "land."

So far, we've explored a Schwarzschild black hole by dropping particles (that is, a person) into it. The next step is to take a look around its neighborhood with a flashlight. This time, suppose that Barbara is safely in the spaceship, and Chris has zipped down to the visible surface of a supermassive star that is about to collapse into a black hole. He is protected from vaporization by his silvery Brin-field spacesuit, and hovers above the maelstrom of plasma on his Niven fusion drive. Chris takes out his flashlight and shines it around to see what happens to the light beam as the star begins its final collapse.

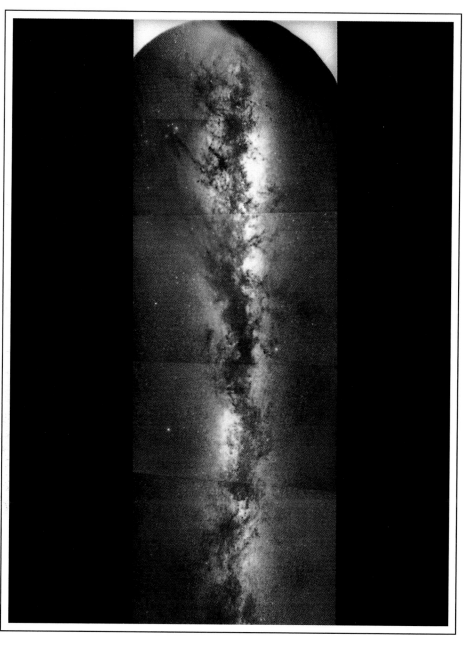

A mosaic of photographs showing the Milky Way stretching from the constellation Sagittarius to Cassiopeia
Photograph courtesy of the California Institute of Technology

The Trifid Nebula (M20) in the constellation Sagittarius is one of the most
spectacular sights in the heavens. The Trifid is a typical emission nebula. Its light
comes from glowing gas excited into visibility by the fierce radiation pouring from
new stars within and around it.
U.S. Naval Observatory

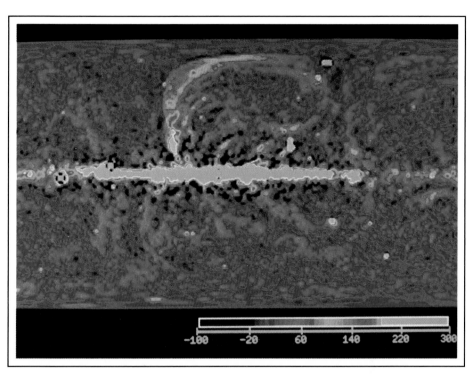

This is how the Milky Way might look if we could see radio waves. This image is of the radio continuum Milky Way.

Image courtesy of Yoshiaki Sofue, University of Tokyo

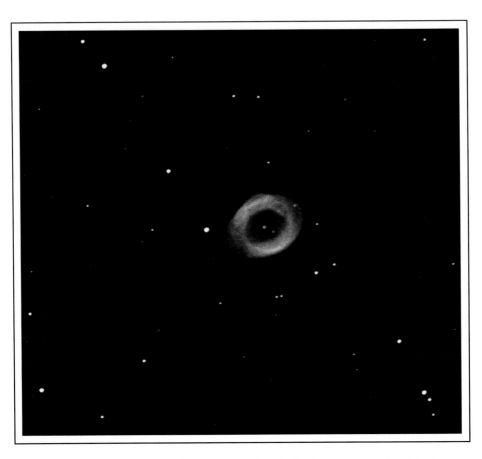

The Ring Nebula (M57) in the constellation Lyra looks like a wedding ring. It is the visible part of a sphere of gas blown out from the aging star at its center.
U.S. Naval Observatory

This is a wide-angle view toward the center of our Galaxy. The teapot shape of the constellation Sagittarius can be seen. However, the central region of the Galaxy is obscured by intervening clouds of gas and dust as well as by the many other stars lying in the line of sight between the Earth and the Galactic core.

National Optical Astronomy Observatories; photograph by Dr. David Talent

The Orion Nebula (M42) is one of the best-known objects in the heavens. This huge cloud of gas and dust is only the visible part of a much larger molecular cloud, the birthplace of many new stars and protostars.
Lick Observatory photograph

A star like the sun and its retinue of planets form from a collapsing cloud of gas and dust within a molecular cloud. Here we see (upper left) the protostar beginning to glow with the light from thermonuclear fusion at the center of the collapsing disk of matter. Next, clumps of dust and gas circling the protostar begin to form. The new star begins to glow more and more brightly, and the veils of gas and dust begin to disperse. Finally (bottom right) the new planets can be seen, their shadows cast out upon the remaining dust and gas that is being blown out of the new solar system by the star's T Tauri wind.

Illustration courtesy of Hansen Planetarium

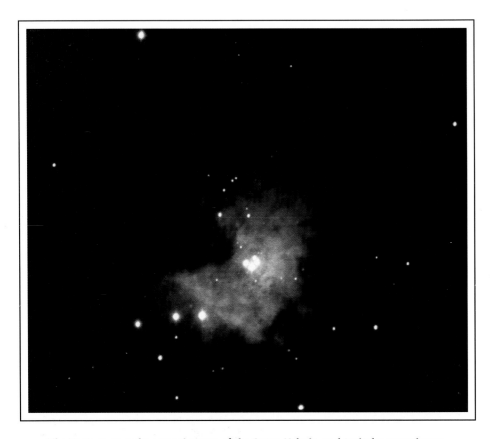

The Trapezium is the central region of the Orion Nebula and includes several very bright, very new stars. These newborn stars are pouring out immense quantities of radiation, which causes the surrounding nebula to glow.

U.S. Naval Observatory

At first Chris notices nothing. Then he observes that the beam of light is being slightly bent by the increasingly powerful gravity field of the star (or, if you wish, by the increasingly distorted space-time around the star). The beam of light still has no problem escaping from the surface of the star. However, it gets more and more bent as the star continues to collapse and the gravity field gets more intense.

Soon Chris notices that when he shines the light beam parallel to the star's surface, the rays are caught in a circular orbit around the star. This region is called the **photon sphere**. The light can still escape from the star when it is shone upward, but not when it is parallel to the surface. Now Chris sees that rays of light pointing at the star's physical horizon are being pulled into the star. Then light beams several degrees from the limb are pulled in.

Chris notes that there is now a cone-shaped region whose axis is perpendicular to the star's surface in which light rays can escape from the star. However, beams of light outside that cone cannot escape. This area is called the **exit cone**, since rays within the cone can still "exit." As the star gets smaller and smaller, and denser and denser, the exit cone gets narrower and narrower. Finally the cone closes. There is no direction in which light rays can travel and still exit the collapsing star. The surface—and Chris and his flashlight with it—have just passed the event horizon. They disappear . . . forever. Theme music from "The Twilight Zone" rises in the background. . . .

Most of what has just happened to our hapless astronauts would take place around any of the four kinds of black holes. A Schwarzschild black hole, however, is one of the two *least likely* kinds to exist. All astronomical objects, including supermassive stars most likely to collapse to a black hole, have some spin on them. An object's rate of spin increases as it becomes smaller, so even a supermassive star with a very slow rotation rate would end up becoming a black hole with at least a moderate amount of spin. Thus, it is extremely unlikely that a nonspinning black hole such as a Schwarzschild black hole would exist in nature. The same can be said of supermassive black holes that might exist at the center of galaxies like our own. Such supermassive black holes would be formed from the collapse and coalescence of thousands—perhaps millions—of stars crowded into the central region of a galaxy. Such stars would be moving very rapidly as they crashed together, and the supermassive object formed from such collisions would almost certainly have some kind of spin. And so would the supermassive black hole that would result.

Still another way to investigate what happens inside a Schwarzschild black hole is with a special kind of space-time diagram that was first developed by physicist Roger Penrose. It is appropriately called a **Penrose diagram**. It has the remarkable property of representing both a black hole and the rest of the entire universe in one fell swoop. Of course, it is a rather simplistic representation of all of space and time, but there it is nonetheless.

This Penrose diagram illustrates a Schwarzschild black hole. All of space-time outside the black hole—that is, the rest of the universe—is represented by the tilted square on the right side of this Penrose diagram. Space and time stretching into the past (bottom) and future (top) extend past the lines at the right edge. At the center of the diagram are two sets of lines inclined at forty-five degrees to the vertical. These are light lines, which represent the black hole's event horizon. The shaded area at the top represents the area of the Schwarzschild black hole that is lying inside the event horizon. That includes the singularity, which is the horizontal jagged line across the top of the Penrose diagram. A singularity travels a spacelike path in space-time.

Any material particle in our universe must follow a timelike path, which is less than forty-five degrees to the vertical. A particle following path *a* remains outside the black hole's event horizon. However, if it follows path *b* it will fall into the event horizon. The only way the particle can avoid encountering the singularity (the line at the top) is by traveling a horizontal spacelike path faster than the speed of light. However, even within the event horizon, such a particle cannot travel faster than light. It must travel a timelike path, so it is doomed to hit the singularity and be destroyed.

This suggests something important about a black hole. Inside the event horizon, the nature of space-time changes. In

Space-Time Diagram for a Schwarzschild Black Hole

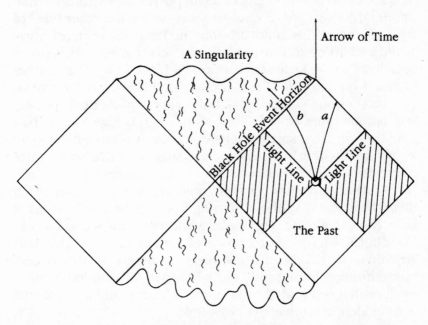

A Penrose diagram for a black hole is a special kind of space-time diagram. Our entire universe is the right-hand tilted box, which contains O (the observer). The two crossing lines at the center represent the event horizon of a Schwarzschild black hole, and the areas to the top and bottom are the singularity at the black hole's center. OA is the worldline of a particle following a path that does not cross the event horizon; it is safe from destruction in the black hole. OB, however, is the worldline of a particle that does cross the event horizon. The only way it can avoid falling into the singularity is if it follows a spacelike path—one inclined at greater than forty-five degrees to the vertical. Since such paths are not permitted in our universe, it is clear that any particle crossing the event horizon of a Schwarzschild black hole will fall into its singularity and be destroyed. The Penrose diagram also reveals the possible existence of another universe on the "other side" of the black hole. However, it can never be reached by something falling into a Schwarzschild black hole.

the outside universe, a particle can move in any direction it wishes, as long as its worldline is timelike. Inside the event horizon, though, its paths are constrained. The only directions it can take are those that lead to the singularity and annihilation.

The Penrose diagram itself suggests still another strangeness about black holes. The diagram possesses a symmetry that implies the existence of *another universe* "on the other side" of the black hole. This is not arbitrary on Penrose's part. Schwarzschild's solutions to Einstein's equations themselves describe a symmetry to the space-time around a black hole. That mathematical symmetry implies a connection between our universe and another one via the black hole. This hypothetical connection is sometimes called an Einstein-Rosen bridge, for the two scientists who first suggested its existence. It is more commonly known as a **wormhole**. Some have suggested that wormholes might connect different parts of our own universe.

If other universes do exist, however, we will never reach them through a wormhole running from a Schwarzschild black hole. Nor can we use wormholes associated with Schwarzschild black holes to travel to other parts of our universe. It is clear that anything (including information) that enters an uncharged, nonspinning Schwarzschild black hole will fall into its singularity. It cannot be avoided without going faster than light, and that is forbidden, even inside a black hole.

The Reissner-Nordström Black Hole

Shortly after Schwarzschild came up with his solution to the general relativity equations, two other physicists published theirs. Hans Reissner in 1916 and Gunnar Nordström in 1918 independently worked out solutions different from those of Schwarzschild. His equations described the gravitational field produced by a nonrotating, uncharged point mass. Nordström's and Reissner's portrayed the gravitational field of a point mass that, though not rotating, does possess an electrical charge. As with the case of Schwarzschild's solutions, scientists later recognized that the two men's solutions described a form of black hole, and it is now named for them.

In some ways a Reissner-Nordström black hole is much like a Schwarzschild black hole. It has no spin, it has only one photon sphere, and its singularity is a point. Also like a Schwarzschild black hole, a Reissner-Nordström black hole is very unlikely to exist in nature. Nearly all astronomical objects do *not* have any net electrical charge. Stars as a whole are electrically neutral. Were they to accumulate a small negative or positive electrical charge (from an excess of electrons or protons, respectively), it would quickly disappear. Positive charges attract negative charges, and vice versa. The same electrical balancing act would occur if an electrically neutral (and nonspinning) black hole were to begin collecting excess electrons or protons and thus accumulate a net electrical charge.

Suppose, though, that a method could be found to add an electrical charge to a Schwarzschild black hole and keep it there, thus creating a Reissner-Nordström black hole. Because it possesses a net electrical charge, its structure would be somewhat different from that of a Schwarzschild black hole. That structure would depend on the amount of charge the Reissner-Nordström black hole carries.

A Schwarzschild black hole has only one event horizon. Suppose that a small electrical charge is now added to the black hole. The event horizon begins to shrink inward toward the singularity, and a *second event horizon* appears still further inward. As the amount of charge on the Reissner-Nordström black hole increases, the outer event horizon continues to shrink, and the inner event horizon begins to expand outward. Put enough electrical charge on the black hole, and the two event horizons merge into one. As the amount of charge increases still further, the now-single event horizon *still* continues to shrink. Eventually the Reissner-Nordström black hole will have a single event horizon lying barely above the singularity. Any more charge added to the black hole will then cause the event horizon to *vanish into the singularity*. The result: the Reissner-Nordström black hole has ceased to be a black hole entirely. It is now a naked singularity, with no event horizon to veil it from the rest of the cosmos. Most physicists devoutly hope that the Reissner-Nordström black hole will remain forever only a theoretical possibility, not a reality.

The Kerr Black Hole

For nearly fifty years the only solutions to Einstein's general relativity equations that portrayed a black hole described one that does not spin. As a simple, idealized model it was fine, but astrophysicists knew that it was hardly a description of something that might actually exist. Any black hole that might really be detected would almost certainly be spinning. The mathematics for describing such an object, however, were incredibly difficult.

In 1963 the difficulties were finally overcome. The journal *Physical Review Letters* published a paper by Roy Kerr, an Australian astronomer working at the University of Texas, entitled "Gravitational Field of a Spinning Mass as an Example of Algebraically Special Metrics." It is a description of what has since become known as a Kerr black hole. This is probably the most likely kind of black hole to exist in nature. It possesses spin but no charge. And because it is spinning, its structure is much more complicated than that of a Schwarzschild black hole or even a Reissner-Nordström black hole.

The spinning of a black hole, first of all, has the same effect as adding a charge. It causes the creation of two event horizons. As spin increases, the inner event horizon moves outward, and the outer one shrinks inward. If the hole spins fast enough, the two will eventually merge, shrink inward to the singularity— and disappear. However, there is still more structure to a spinning black hole than two (possible) event horizons.

When an object like a black hole is spinning, other objects that are nearby will tend to be dragged along in the direction of its spin. This phenomenon is called **frame dragging**, and it is connected with an idea first proposed by the physicist Ernst Mach (1838–1916) back in 1872. Mach believed that the property of matter we call inertia (which, as defined earlier, is matter's "laziness," its tendency to stay in one place unless acted upon) is not intrinsic to matter. Rather, he argued, it arises from some kind of mysterious interaction between one individual object and all the other mass in the universe. If the Earth were the only object in the universe, it would have no inertia. It is only the presence of all the other matter in the cosmos that confers inertia on any particular object.

Newton had argued that inertia is conferred upon matter by what he called absolute space. Mach was rejecting Newton's idea of absolute space. Einstein was later quite impressed with Mach's proposal (not surprisingly, since he would also reject absolute space and absolute time), and he dubbed it **Mach's principle**.

Frame dragging has to do with Mach's principle. Since most of the matter in the universe is always very distant from any one particular object, it is the most distant matter that contributes to inertia. However, matter that is nearby also has a tiny but real effect. That being the case, a rotating body that is extremely massive—namely, a black hole—would have the effect of "dragging along" the inertial frames of reference of any other objects close by. The effect of frame dragging is somewhat analogous to the way water is spun into a whirlpool. We would not notice the frame-dragging effect if we were standing next to, say, a spinning model of the Earth in the local library. The globe's mass is so tiny that the frame-dragging effect is infinitesimal. However, if we were near a spinning white dwarf star, or neutron star, or black hole, we would most certainly feel the effects of frame dragging.

Suppose our intrepid black-hole explorers, Barbara and Chris, are in their spaceship and approaching a Kerr black hole. Frame dragging will cause the ship to be pulled along in space in the direction of the hole's spin. As long as they remain a considerable distance from the hole (a distance that depends on the black hole's mass and rate of spin), they will be able to use the ship's engines to counteract the pull of the frame-dragging effect and remain motionless (relative to the distant stars, of course) above the hole. The closer they get to the Kerr black hole, the more thrust they will need to counteract frame dragging.

Finally they reach a point where frame dragging will pull them around the hole, no matter how powerful their engines may be. This point is called the **static limit**; it is the region surrounding a spinning black hole where a particle cannot remain at rest. Even if Chris and Barbara fall below the black hole's static limit, they are not permanently trapped; they have not yet fallen below the hole's event horizon. Their spaceship uses a Forward/Davis antimatter propulsion system, which is powerful enough to enable them to still escape from the black hole.

The static limit, like the event horizon, can be pictured as an ovoid sphere surrounding the spinning black hole. The static limit and the event horizon touch at the poles of the black hole. The greatest separation between them is at the equator. This region between the static limit and the event horizon is called the **ergosphere**. The name is carefully chosen: *ergo-* is a Greek-based prefix meaning "work" or "energy."

In 1969 Roger Penrose showed that it is possible to *extract energy from a black hole*, specifically from this region called the ergosphere. Suppose Chris and Barbara's spaceship is a two-stage vehicle. As soon as they plunge beneath the static limit into the ergosphere, they jettison the first stage, dropping it into the singularity. The second stage (with them in it, of course) emerges from the static limit. This second, escaping part of the ship will leave the black hole with much more energy than it had before. The energy has to come from somewhere, of course; Heinlein's law clearly states, "There ain't no such thing as a free lunch." (See Robert Heinlein's award-winning science fiction novel *The Moon Is a Harsh Mistress* for a delightful and sobering explanation of this law, abbreviated as TANSTAAFL.) The ship's additional energy is the energy of the black hole's spin. The hole is now spinning a bit more slowly than before.

There is a limit to how much energy we can extract from a black hole. That limit is reached when the black hole's spin reaches zero. If we begin extracting energy from a Kerr black hole spinning at its greatest possible rate (called a maximal Kerr black hole), we can extract only 29 percent of the hole's initial mass-energy. However, that's quite a bit of energy. The thermonuclear fusion processes that power stars turn only about 1 percent of a star's mass into energy. Clearly, spinning black holes could be a source of immense energy for any civilization that could harness them. And they may already be the most powerful energy sources in the universe.

All event horizons around spinning black holes have static limits, so a Kerr black hole can have two static limits (one for each event horizon) and two ergospheres. As the spin of a Kerr black hole increases, the two static limits move closer together and finally merge. Both the static limits and event horizons (now merged into one of each) move closer to the singularity and finally disappear into it, leaving a naked singularity. However, the singularity of a Kerr black hole is different from that of a Schwarzschild black hole. Because the hole is spinning, the singularity is not a point but a *ring* singularity.

Finally, there is the matter of the photon sphere. As with the case of event horizons and static limits, a spinning black hole has two photon spheres. The inner one is closer to the singularity than the photon sphere of a Schwarzschild black hole. The outer one is slightly farther away. The faster the hole spins, the farther from each other the two photon spheres move.

The two do differ from each other. A beam of light moving in a direction opposite to that of the black hole's spin will be caught in the outer photon sphere. If the light beam is shone in the same direction as the hole's spin, it will not be captured in orbit around the hole until it is brought to the distance of the inner photon sphere. In other words, if the black hole has a clockwise rotation, a light beam in the outer photon sphere will be moving counterclockwise; one in the inner photon sphere will be moving clockwise.

The Kerr-Newman Black Hole

Two years after Roy Kerr published his description of a spinning black hole, physicist E. T. Newman and five colleagues came out with a mathematical depiction of a spinning *charged* black hole. The Kerr-Newman solutions to the black hole equations appear to cover all the possible types of black holes. A Kerr-Newman black hole is less likely to exist in nature than a Kerr black hole, but because it has spin, it is more likely to exist than either a Schwarzschild black hole or Reissner-Nordström black hole.

The structure of a Kerr-Newman black hole is like that of a Kerr black hole. It can have two event horizons, two static limits, and two ergospheres between them. It can have two photon spheres, and it contains a ring singularity. Finally, unlike a Kerr black hole but like a Reissner-Nordström black hole, a Kerr-Newman black hole also has an electrical charge.

We've already examined the interior of a nonspinning uncharged black hole. What would the interior of a "more realistic" spinning Kerr or Kerr-Newman black hole look like? It would be considerably different from that of a Schwarzschild black hole. The singularity, as we've seen, would have the form of a ring instead of a point. If it is plotted on a Penrose diagram, it turns out to be vertical instead of horizontal. The singularity of a spinning black hole seems to be timelike instead of spacelike. The Penrose diagram for a spinning black hole, as shown here, must include both its outer and inner event horizons.

When we plot the different worldlines of *a* and *b*, we find something quite intriguing. It appears there is a way to avoid hitting the singularity of a spinning black hole. An object falling through the outer event horizon along path *a* will eventually encounter the singularity and be destroyed. This path corresponds only to objects that fall into the Kerr black hole from its equatorial plane. However, an object falling along path *b avoids* the singularity.

The reason is that something amazing appears to happen when an object passes through the *inner* event horizon. Within the outer event horizon (as we saw earlier), the properties of space-time change in such a way that an object cannot move in a direction that takes it away from the singularity. However, within the inner event horizon, the properties of space-time change back again. The object is again able to move in directions away from the singularity.

Suppose our heroic astronauts, Chris and Barbara, have fallen in their spaceship through the inner event horizon of a spinning black hole. They fire up their Forward/Davis antimatter propulsion system and alter their course away from the ring singularity. They are able to do this without exceeding the speed of light, thus traveling on a timelike path, which on the Penrose diagram is inclined by less than forty-five degrees to the vertical axis. Their path misses the vertical line of the singularity, passes through *another* set of inner and outer event horizons, and takes them out of the black hole—*into another universe*. Once again, the wormhole makes its appearance in a black hole. Unlike the case of a Schwarzschild black hole, however, it appears to be possible for this bridge to another universe to exist with spinning black holes.

Space-Time Diagram for a Spinning Black Hole

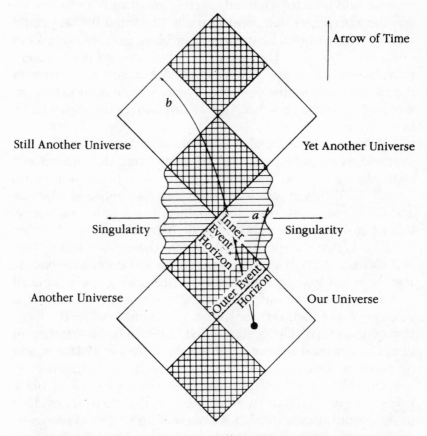

A spinning or Kerr black hole has two event horizons, outer and inner. Space-time
between the two event horizons is different in nature from that beyond the inner
event horizon. In the former, an object cannot move in a direction taking it away
from the singularity. In the latter, space-time becomes like our space-time once
more, and an object is again free to move away from the singularity. A Penrose
diagram for a Kerr black hole shows that it is theoretically possible to travel through
the black hole and not be destroyed by the singularity. The object traveling on
worldline a falls in from the Kerr black hole's equatorial region and passes through
the outer and inner event horizons, but its path takes it into the ringlike singularity,
where it is obliterated. The object traveling on worldline b passes through the outer
and inner event horizons but is able to maneuver within the inner event horizon.
It eventually emerges in "still another universe." Following a slightly different path,
it could also emerge in "yet another universe." However, it could not travel to
"another universe," since that would require that it follow a spacelike path through
the outer event horizon at a velocity faster than light.

It would appear, in fact, that an infinite number of universes could be daisy-chained together by spinning black holes and the wormholes that connect them. Chris and Barbara could merrily pass through black hole after black hole, moving from universe to universe. They would only be moving from a present universe into a "future" universe, of course, since the arrow of time moves only in that direction. There is no indication that the arrow of time changes direction as one moves through a black hole.

However, there are also a couple of variations on this. For example, some scientists have suggested that the "universes" charted in a Penrose diagram of a spinning black hole could actually be different locations within our own universe. Barbara and Chris could use the daisy chain of black holes as "gates" that open into future versions of their own universe. Another variation is that the spinning-black-hole daisy chain leads Chris and Barbara through a series of universes and *eventually back to their own universe,* just as a Möbius strip bends back upon itself and ends up having only one side. They might even reemerge *into the past* of their own universe. This would violate the basic law of causality, which states that cause must come before effect, and would appear to render the universe a place where no one can make scientific predictions with any confidence.

One objection to all this universe hopping via Kerr black holes is pretty straightforward. Any object that even comes close to the event horizon of a black hole will be ripped to pieces by the immense tidal effects. However, this would be the case only with a stellar-mass black hole. As noted earlier, supermassive black holes with the mass of several hundred million or even billion suns do not have appreciable tidal effects at their event horizon. Chris and Barbara could cruise through the outer and inner event horizons of such a black hole in their spaceship and encounter no tidal effects worth mentioning.

There is, however, a more serious objection to spinning black holes as "gates" into other universes and/or the past. As complex as the equations for Kerr black holes and Kerr-Newman black holes may be, the pictures we have developed of their interiors appear to be somewhat idealized and simplistic. These models of black holes, among other things, ignore the effects of quantum physics. We've already seen that quantum effects might make it impossible for a singularity ever to exist. It also appears that quantum effects would destroy any wormhole that developed in a charged (Reissner-Nordström) black hole. Charged black holes and spinning black holes have similar internal structures, so it's likely that the same quantum effects also would wipe out a wormhole in a spinning (Kerr or Kerr-Newman) black hole.

Nevertheless, it would still be possible to use a supermassive black hole as a gate into the *future*. Remember that time outside a black hole speeds up for someone who is falling toward the event horizon. Remember, too, that it is possible for someone in a spaceship to travel into the ergosphere of a supermassive spinning black hole and still come out in one piece.

Suppose Chris and Barbara do just that. At the center of the Galaxy, they take their spaceship past the static limit of the supermassive black hole and travel into its ergosphere. They then fire their Forward/Davis antimatter engines and come back out, past the static limit and into normal space. While they may have spent only a few seconds *inside* the black hole's ergosphere by their frame of reference, *outside* in the Galaxy at large, several million years will have passed. They have effectively— and permanently—traveled into the future.

Soon we will, too.

PART III
A JOURNEY THROUGH SPACE

8

STAR LIGHT, STAR BRIGHT

Like boys and girls in the Mother Goose rhyme, galaxies are made of many things. In this case, it's not sugar and spice, or snails and puppy dog tails, but stars and gas and dust, as well as something still mysterious and unseen called "dark matter." What is visible, though, are the stars. Galaxies are star cities, stellar metropolises so vast that the light of a star at one end of the city can take hundreds of thousands of years to get to the other side. Like a living creature, a galaxy's bones, muscles, and organs may lie hidden, but its skin of suns is quite visible. We cannot understand galaxies in general and our own Galaxy in particular without having at least a little knowledge of stars— their origin, composition, types, lifestyles, and ultimate fate.

Stars have many different properties, and astronomers have devised several different ways of measuring them. Perhaps the most obvious characteristics to the naked eye are brightness and color. Astronomers have also come up with ways to measure stars' size, temperature, chemical composition, and velocities, among other things. All these stellar properties are important for understanding the origin and evolution of the Galaxy and what may be happening at its center.

Stellar Brightness

Astronomers usually measure a star's brightness in terms of "apparent magnitude" and "absolute magnitude." The magnitude scale is a somewhat quirky one, largely because of its historical origin. The great Greek philosopher/scientist Hipparchus first classified stars by their apparent magnitude. **Apparent magnitude** is the brightness of a star as it is seen from Earth. Hipparchus rated the brightest star he could see as magnitude 1 and the faintest as magnitude 6. Over the centuries astronomers realized that some stars were brighter than Hipparchus's magnitude 1 stars, so the scale has extended past zero into negative numbers. The star Vega, for example, is apparent magnitude 0. Sirius is magnitude −1.4.

Also note that the scale is upside down: the *larger* the number, the *fainter* the star. Thus, a magnitude 7.5 star is fainter, not brighter, than a star of magnitude 5.5, and a star with a magnitude 10.1 is brighter than one of magnitude 12.6. One way to avoid the mental confusion is to regard magnitude as a measure of *faintness* instead of brightness.

Another peculiar aspect of the magnitude scale is the brightness (or faintness) ratio between numbers. Hipparchus considered a star of magnitude 1 to be a hundred times as bright as one of magnitude 6, a brightness ratio of 100 for a difference of five magnitudes. Therefore, in today's magnitude scale, a difference of one magnitude is a brightness ratio of 2.512. This seems like a strange ratio, and it is. But it makes at least mathematical sense, because 2.512^5 (or $2.512 \times 2.512 \times 2.512 \times 2.512 \times 2.512$) = 100, the brightness ratio for a difference of five (1 to 2, 2 to 3, 3 to 4, 4 to 5, 5 to 6) magnitudes. The accompanying table may help to keep it all straight.

In a more practical vein, the sun has an apparent magnitude of about −27, while the faintest stars visible to the naked eye are at about magnitude 6. Using binoculars, we can see stars as faint as magnitude 10.

Astronomers also use **absolute magnitude**. This is the brightness of a star if it were at a preset standard distance from Earth. That standard distance is defined as 10 parsecs, or 32.6 light-years. Stars that are actually closer to us than this distance would appear fainter at 10 parsecs. Those that are farther away (which is all but a few dozen or so) would appear brighter than they actually do. The brightness of a star as it would appear at 10 parsecs from us is its absolute magnitude.

Apparent magnitude and absolute magnitude are usually

CONVERTING MAGNITUDE DIFFERENCES INTO BRIGHTNESS RATIOS

Magnitude Difference of	=	Brightness Ratio	Which Is About—
0.0		1.0	
0.1		1.1	$2.512^{0.1}$
0.2		1.2	$2.512^{0.2}$
1.0		2.5	
1.5		4.0	$2.512^{1.5}$
2.0		6.3	2.512^{2}
2.5		10.0	$2.512^{2.5}$
4.0		40.0	2.512^{4}
5.0		100.0	2.512^{5}
10.0		10,000.0	2.512^{10}
20.0		100,000,000.0	2.512^{20}

quite different. For example, the sun's apparent magnitude is −26.7; it is only eight light-minutes away, of course. The apparent magnitude of the star Sirius is −1.46. However, the sun's *absolute* magnitude is 4.1, while that of Sirius is 1.4. The difference is about 2.7 magnitudes, so at visual wavelengths, Sirius is actually twenty-two times as luminous ($2.512^{3.4}$) as the sun.

Apparent and absolute magnitudes are measurements of luminosity only at visual wavelengths, red through violet. **Bolometric magnitude** is the magnitude of a star measured through all its wavelengths. Distance and apparent and absolute magnitude are clearly related. If we know two of these properties, we can figure out the third.

Spectral Classes

The spectra of most stars resemble the spectrum of our own sun, so astronomers are able to determine the elements in a star's photosphere by the absorption lines in its spectrum. However, there is an important additional complication. A star's surface temperature as well as its composition will affect its spectrum. For example, the Balmer lines in the spectrum are fainter and weaker in cooler stars and broader and darker in hotter stars—up to a point. In very hot stars the Balmer lines again become weak. The surface temperatures are so high that the electrons are completely stripped away from many hydrogen nuclei. That leaves few hydrogen atoms with electrons that are jumping into and out of their second energy level, and thus faint Balmer lines.

At the beginning of this century, some astronomers and technical workers at Harvard College Observatory began classifying the spectra of stars by comparing their absorption lines, in particular the hydrogen Balmer lines. The most important and extensive work was done by Annie Jump Cannon. This unsung hero of modern astronomy (she was not a "professional" astronomer, but a "mere" technical worker—and a "mere" woman to boot) classified the spectra of over a quarter million stars.

Originally the classification was strictly alphabetical—class A for stars with the strongest Balmer lines, class B for those with slightly weaker lines, and so on. As the years passed, some classes were dropped, and the order was rearranged into one of decreasing stellar temperature. The classification scheme today runs O-B-A-F-G-K-M. The standard mnemonic for remembering the sequence is "Oh, Be A Fine Girl/Guy, Kiss Me!" In recent years astronomers have added two other classes to the sequence, C for carbon stars and S for heavy-metal stars with temperatures similar to those of K and M stars. Perhaps the phrase "—Come on, Sweetie!" could be added to the classic mnemonic to update it. The accompanying table lists the seven major spectral classes of stars, along with some of their features.

These spectral classes are further subdivided into ten subclasses with the numbers 0 to 9 placed after the spectral type letter. So G5 is midway between G0 and K0. Other characteristics of the star's spectrum can be denoted by adding an additional lowercase letter. For example, e means that emission lines are present in the stellar spectrum, m that metallic lines are present, p that it's a peculiar spectrum, and so on.

In the 1930s astronomers at Yerkes Observatory developed a classification system based on the brightness or luminosity of a star. It is called the **MK system**, named for the two astrono-

FEATURES OF THE STELLAR SPECTRAL CLASSES

Class	Color	Temperature (K)	Examples
O	Bluish white	30,000–50,000	Naos (in the constellation Puppis)
B	Bluish white	11,000–25,000	Rigel, Spica
A	Bluish white	7,500–11,000	Sirius, Vega
F	Bluish white to white	6,000–7,500	Canopus, Procyon
G	White to yellowish white	5,000–6,000	Capella, the sun
K	Yellowish orange	3,500–5,000	Arcturus, Aldebaran
M	Reddish	3,500	Betelgeuse, Antares

mers who developed it (Morgan and Keenan). The MK system has six luminosity classes, denoted by the Roman numerals I through VI: I is a supergiant star, II a bright giant star, III a giant star, IV a subgiant star, V a dwarf star, and VI a subdwarf star. White dwarf stars are now put into a seventh class named VII. The luminosity class number is placed after the spectral letter and its subclass numbers and letters.

Using these classification systems, the sun is classed as a G2 V star: a dwarf star of spectral class G2, white to yellowish white in color, with a surface temperature of 5,780 K (above absolute zero; helium becomes liquid at 4.2 K; water ice melts at 273.15 K; the center of the sun is about 10^6 K).

Temperature, Diameter, Chemical Composition

The temperature of a star is not something we can measure directly. No one can travel to a star, insert a thermometer into it, and measure its temperature. However, there are still ways to determine this important stellar property. Earlier we came across the concept of blackbodies and the way Max Planck solved a pressing theoretical problem involving them and ended up becoming the reluctant grandfather of quantum physics. Blackbodies are important to astronomy, because the sun and the stars radiate energy very much as a blackbody does. There are two important mathematical equations related to blackbodies which astronomers use to measure the temperature of stars. One is called the Stefan-Boltzmann law, and the other is called Wien's law. The former can be used to measure stellar temperature if we know how much energy each square meter of the star is putting out. If we know the peak or strongest wavelength of a star's spectrum, we can use the second equation.

We can also measure the **color temperature** of a star by measuring its relative brightness at any two wavelengths. That in turn can give us an estimate of a star's size. Suppose we are examining two stars. One is a red star with a surface temperature of about 3,000 K, and the other is a bluish star with a surface temperature of around 15,000 K. Both are the same distance from the sun. However, the first star is ten times as bright as the second. This apparent paradox is resolved when we remember that stars radiate energy as blackbodies do. The amount of energy emitted by each unit of area of a blackbody depends only on the fourth power of its temperature, nothing else. The second star is five times hotter than the first, so each square meter of its surface emits 5^4 or 625 times as much energy as the first star. If the first star had 625 times the surface area of the second, both would have the same luminosity. But since it's ten times more luminous, it must have 10×625 or 6,250 times the surface area of the second star. That in turn means it has a radius some twenty-two times as large as the second star.

It's clear that a star's brightness is related not only to its temperature but also to its size. This gives us a way to determine a star's relative surface area and radius. If we can make comparisons with a star whose radius and luminosity we know accurately—that is, the sun—we can determine a star's size relative to the sun.

We can use another way to measure the diameter of a star if we know its distance and its **angular diameter**. The angular diameter of a celestial body is the observed diameter it covers in the sky expressed in degrees, minutes, or seconds of arc. Stars are so distant from us that their angular diameter is usually stated in milli–arc seconds.

Ground-based telescopes cannot directly resolve the diameter of a star because our always-turbulent atmosphere blurs the stellar image. However, recent techniques are beginning to find ways around this problem, and the Hubble Space Telescope may eventually be able to directly image the disks of nearby stars. Meanwhile, it has still been possible to determine some angular diameters for stars by using an instrument called a stellar interferometer. Another way is by watching lunar occultations, when the moon passes in front of a star. By accurately determining the time it takes for a star's light to be blocked by the moving moon, one can determine the star's diameter. This works only if we know how distant the star is from us, but it has been successfully carried out. One example is of the star Aldebaran, the red eye of the constellation Taurus, the Bull. Aldebaran is known to be about sixty-four light-years away. Lunar occultations give a figure of about twenty-one milli–arc seconds for its angular diameter, which in this case works out to a radius of about 32.7 million kilometers, some forty-seven times greater than the sun.

The elements that are present in a star's surface layer can be determined from its spectrum, as we've seen earlier. Theoretical and observational evidence makes it clear that nearly all stars are made almost entirely of hydrogen and helium. It is only when stars have reached the ends of their lives that they have turned even a small fraction of their hydrogen into heavier elements like magnesium and iron.

The Color of Stars

Color is another distinctive property of stars, even when we are looking at them with our unaided eyes. Betelgeuse looks red, Capella is yellow, Sirius is white, Rigel is blue-white. A star's color is related to its temperature, something easy to understand. Were we to take a bar of cold iron and place it in a very hot fire, it would get hotter and hotter. As the temperature of the iron bar rose, we would first see it start to glow a dull red, then orange, then yellow, and finally white. In the same way, the hotter the surface temperature of a star, the brighter its color.

Of course, we only see the visible-light wavelengths of starlight, not all of the radiation they emit. Stars radiate at many different wavelengths, more strongly at some than at others. So one way astronomers define stellar "color" is to measure the difference between a star's relative brightness at two standard wavelengths. A blue star emits more than just blue light, but it is brighter at the blue wavelengths than at the red ones. This difference measurement is standardized in several systems called **color indexes**.

The value of a star's color index depends on whether it is mainly blue, red, yellow, and so on. Because it's an indication of the star's color, the index also indicates its temperature. The color index is completely independent of a star's distance from the Earth.

Astronomers today mostly use the UBV system, in which the color index is the difference between B and V. B is blue starlight at a wavelength of 440 nm, and V is greenish yellow starlight at 550 nm. Sometimes the U-B color index is used, in which U is an ultraviolet wavelength of 365 nm—thus the name UBV for the system.

Each spectral class of stars (O, B, A, F, G, K, M) has a characteristic intrinsic color index, designated as $(B\text{-}V)_0$ and $(U\text{-}B)_0$. Both $(B\text{-}V)_0$ and $(U\text{-}B)_0$ are defined as zero for class A0 main sequence stars, so hotter stars such as O and B stars have a negative intrinsic color index, and cooler A1 through M stars have a positive one.

Color indexes also exist for magnitudes at red and infrared wavelengths. Two examples are the V-R and V-I indexes. R is the star's apparent magnitude measured at a wavelength of 0.7 μm, and I is the magnitude at 0.9 μm.

The H-R Diagram

Astronomers, like most other scientists, have a penchant for classifying the objects they study. They classify because they are looking for patterns, since patterns can reveal new information about what is studied. The Austrian monk Gregor Mendel saw patterns in the colors of the peas he was growing in his garden, and he discovered the laws of heredity. Charles Darwin espied patterns in the shapes and sizes of birds' bills and eventually developed the theory of evolution. Spectral classes and color indexes are examples of ways in which astronomers classify stars and uncover new patterns. In the early part of this century, two astronomers found patterns in stellar temperature and luminosity, and they developed an important tool for studying stars.

Ejnar Hertzsprung was a Danish astronomer, and Henry N. Russell was an American astronomer. Independently of each other, the two men began categorizing stars by their spectral type (and thus temperature) and brightness. Hertzsprung in 1911 and Russell in 1913 discovered something interesting when they turned their information into a graph. Each star was represented on the graph by a point that denoted its absolute brightness and its temperature or spectral class. As data accumulated on their respective graphs, a pattern began to emerge. Stars were not distributed uniformly across the graph, but rather fell into well-defined bands or groups. The type of graph that these two astronomers independently developed is now called a **Hertzsprung-Russell diagram**, or **H-R diagram**.

About 90 percent of all stars fall into a diagonal band that runs from the upper left to the lower right of the classic H-R diagram. This band is called the **main sequence**. The sun lies on the main sequence, slightly below and to the right of the diagonal's midpoint. Stars that are somewhat larger and brighter than main sequence stars lie in a broad band that stretches off to the right from the main sequence's middle. Extremely bright and rare supergiant stars lie in a diffuse cluster above the giant band. White dwarf stars lie in a scattered group at the bottom left of the H-R diagram.

Several different forms of H-R diagrams can be constructed using different sets of information. For most astronomical applications, absolute visual magnitude is plotted on the vertical axis, and spectral type, temperature, or color index on the horizontal axis. Because a star's color index is very easy to measure, it is usually used in H-R diagrams instead of spectral type or temperature. Such diagrams are called **color-magnitude H-R diagrams**. Other kinds include **color-luminosity H-R diagrams** and **theoretical H-R diagrams**.

H-R diagrams have found many uses in astronomy. One can plot theoretical H-R diagrams and then test the theories by comparing the diagrams with others based on real-life observations. The diagrams can also be used to determine distances to some stars. And in 1944 an astronomer used H-R diagrams to discover the existence of two different populations of stars. That discovery had a direct impact on our increasing understanding of the nature of our own Galaxy.

Stellar Populations and Age

Stars are not all the same age, but that isn't immediately apparent from looking at the night sky. Astronomers came to realize this from careful examination of stellar spectra and from an increasing understanding of the origin, evolution, and age of the entire universe.

In the 1930s and 1940s astronomers noticed an interesting difference between stars found in two different kinds of groupings, *globular clusters* and *open clusters*. Most open clusters (sometimes called galactic clusters) are found in or near the Galactic plane. A typical open cluster is the Pleiades, in the constellation Taurus ("The Bull"). This cluster of stars is about four hundred light-years from Earth. The Pleiades were known to the ancient Romans and Greeks as "The Seven Sisters," since seven (today six) stars in the cluster were visible to the naked eye. In fact, the Pleiades contain about one hundred stars surrounded by wisps of gas. The typical open cluster has anywhere from a hundred to a thousand stars in an area of a few tens of light-years. The Pleiades are a pretty average open cluster. Astronomers discovered that most of the stars in the Pleiades below class A0 and above class M0 fall almost squarely on the main sequence line of an H-R diagram.

Globular clusters are different from open clusters. One good example is the globular cluster 47 Tucanae, found in the southern constellation Tucanae ("The Toucan"). Globular clusters are spherical cities of stars that circle the Galaxy in highly eccentric orbits taking them as far as a hundred thousand light-years from the Galactic center. Globular clusters have anywhere from ten thousand to a million stars crammed into spheres with diameters ranging from thirty to three hundred light-years. The stellar density increases toward the center of the globular cluster, where the density can be as high as thirty stars per cubic light-year.

The H-R diagram of a globular cluster is quite different from that of an open cluster. Many of the stars turn off from the main sequence diagonal and angle up to the region of giant stars. Even more striking is a so-called "horizontal branch" of stars, which stretches from the red-giant branch back across to the top of the main sequence line. The horizontal branch on an H-R diagram is the signature of a globular cluster.

It was clear from the H-R diagrams that the stars in open clusters and in globular clusters were of very different types. Visual and photographic examination also presented obvious differences. The stars in open clusters tend to be blue-white. Those in globular clusters tend to be more reddish.

Bearing in mind these distinctions between open cluster and globular cluster stars, the astronomer Walter Baade in 1944 proposed the existence of two distinct populations of stars. These classifications are called **Population I** and **Population II**. Baade proposed that the blue-white stars in open clusters be called Population I stars, while the redder and less luminous stars in globular clusters be called Population II stars.

Meanwhile, an even more interesting distinction was showing up in the chemical composition of the two populations. Spectroscopic examination revealed that Population I stars have pretty much the same chemical makeup as the sun. About 1 to 2 percent by mass is elements heavier than hydrogen and helium. Population II stars, by contrast, contain only 0.01 to 0.02 percent heavy elements. This difference in the ratio of heavy elements to hydrogen and helium is expressed by astronomers as the **[Fe/H] ratio**.

All of the hydrogen (which has the chemical abbreviation H) and helium (or He) was created during the first few minutes of the Big Bang, at the beginning of space-time and matter/energy. Almost all the rest of the elements were created later. Some elements were "cooked" by fusion reactions in the interiors of stars, while other elements were created in the cataclysmic instants of supernova explosions. These heavier elements were formed in more or less even amounts.

Iron, abbreviated as Fe by chemists, is the heaviest element that can be cooked up in the interior of a nonexploding star. It is also one that is easy for astronomers to detect and measure in the optical part of stellar spectra. By measuring the abundance of iron in a star, an astronomer is basically measuring the abundance of *all* the star's heavy elements. Of course, this is not entirely true all the time, because stars often do funny things in their interiors. To a great extent, however, a star's abundance of iron is an indication of its abundance of heavy elements.

After determining the abundance of heavy elements in a star, astronomers can compare this amount with the known abundance of primordial hydrogen and helium. This is expressed as [Fe/H]—the abundance of a star's heavy elements (represented by "Fe") compared to its hydrogen and helium (represented by "H"). If astronomers used only the actual [Fe/H] of a star, they would be working with some extremely small numbers. The brackets that surround the expression [Fe/H] add a key part to the ratio. They mean that this ratio is to be compared *with that of the sun* as a standard.

The actual [Fe/H] ratio is expressed as a positive or negative number. It is often a decimal, and it is also logarithmic. For example, a star whose [Fe/H] ratio is 0 has a ratio of heavy elements to hydrogen that is the same as for the sun. If [Fe/H] is −1, it is *one-tenth* of the sun's ratio. An [Fe/H] ratio of −2 is *one-hundredth* that of the sun. Stars that have positive [Fe/H] ratios are called **metal-strong**. Those with negative ratios are called **metal-weak** or **metal-poor**.

Most astronomers used to consider metal-weak stars to be very old. They contain relatively little heavy elements because they were most likely formed early in the universe's history, when there was only a small amount of heavy elements in the interstellar medium. Today it is also known that some metal-weak stars can be younger stars that formed in an environment that has had very little elemental processing. For example, in several nearby dwarf galaxies, the young stars are metal-poor. This indicates that these galaxies have a very slow rate of stellar evolution processes.

For our own Galaxy the metallicity ratio is still a fairly good indicator of a star's relative age. In other galaxies it is not. However, our Galaxy is massive, quite active in terms of star formation, and fairly coherent in structure. So astronomers continue to confidently hold that a star with a low metallicity ratio is probably an old star.

The significance of the difference in stars' chemical composition became clear with our greater understanding of the origin and evolution of the universe. Edwin Hubble's discovery of the cosmic redshift in 1929 was the first clear observational evidence that the universe is not static. It had a beginning, which today is usually referred to as the Big Bang. The current best estimate is that the universe—and therefore all of time and space—began about eighteen billion years ago. Nearly all the matter in the universe is hydrogen (73 percent by mass) and helium (25 percent). All the hydrogen and all but 1 percent of the helium were formed in the early phases of the Big Bang. The rest of the helium and all the other elements were formed in the thermonuclear furnaces that are the interiors of stars.

The earliest stars to form in the universe must have been made of nothing but hydrogen, along with small amounts of helium. Heavier elements like carbon, oxygen, neon, and iron were formed by fusion reactions at their centers. Elements heavier than iron, such as silver, gold, and uranium, were created in the fierce fires of supernovae, as many of these early stars exploded at the ends of their lives. New generations of stars later formed, and included in their gas and dust were traces of those heavier elements.

Population II stars must be older than Population I stars. They have fewer heavier elements because they were formed when the universe was young, and there were no (or only very small amounts of) heavy elements. By contrast, Population I stars have more heavy elements in their chemical composition because they formed from material enriched in heavy elements and recycled from the remains of older exploded stars.

Astronomers therefore came to identify Population I stars as the young, relatively metal-rich, and highly luminous stars found in the Galaxy's spiral arms. They are often to be found in open clusters, as well as in and around clouds of gas and dust called nebulas. Population II stars, on the other hand, are old and red, and ten to a hundred times less luminous than Population I stars. They are found mainly in the globular clusters, as faint stars in the Galactic halo, and in the Galactic bulge that surrounds the center of the Galaxy.

In recent years astronomers have come to realize that the distinction between Population I and II stars is considerably more fuzzy than Walter Baade first thought. Not all Population I stars are luminous and blue-white. Many Population I stars are stars like our sun, yellowish in color and average in luminosity. The vast majority of Population I stars are faint red dwarfs. By contrast, not all red giants are Population II stars.

The truth is that there is a continuum of "populations" that stretches between Baade's Population I and Population II stars. The Galaxy's earliest and oldest stars are sometimes referred to as **halo Population II stars**. These are the ones found in the Galaxy's extended halo, including globular cluster stars. The Galaxy's oldest known stars are found in the globular cluster named M3, which has an age of 12 to 14 billion years. Then come **intermediate Population II stars** found in the Galaxy's disk, followed by the **disk Population stars**. These may be young Population II or old Population I stars. The sun is about 5.5 billion years old. The open cluster named M67 is about 3.2 billion years old. Successive generations of Population I stars have formed in a series of star formation episodes throughout the Galaxy. The youngest stars, sometimes called **extreme Population I stars**, have twenty to thirty times the abundance of heavy elements of the oldest halo Population II stars. The Hyades, another open cluster in Taurus, contains stars no more than 700 million years old. The double cluster h and Chi Persei is only about 10 million years old and contains some of the Galaxy's youngest stars and protostars.

Finally, some astronomers have suggested the existence of a third population of stars, called **Population III**. This population of supermassive stars could have existed even before the galaxies began to form. However, Population III stars are purely hypothetical. There is no firm evidence that they exist or ever existed.

Variable Stars

A variable star is one whose physical properties change with time. The easiest variation to detect for most stars is brightness, so the term *variable star* usually is applied to stars whose brightness changes over time. The change in brightness can be regular or irregular, and the periods of variable stars range from a few minutes to many years. The amplitude, or range from maximum to minimum brightness, also differs considerably from one variable star to another. Some have amplitudes that are barely perceptible. Others have amplitudes that can be seen by the naked eye.

Astronomers group variable stars into three major categories: cataclysmic, eclipsing, and pulsating. Other, more minor classes of variable stars include flare stars, rotating variables, and spectrum variables. It has been suggested, without too much evidence, that our own sun may be an irregular variable or a minor flare star.

Cataclysmic Variable Stars

A cataclysmic variable is a double star (two stars orbiting each other) where one star is a white dwarf that is having matter dumped onto it from the other component. Eventually the star-stuff accumulating on the white dwarf reaches the point where it begins to flare up and cause a sudden and unpredictable change in brightness. The best known cataclysmic variables are the stars called novae. (Novae should not be confused with supernovae, which are entirely different kinds of stars undergoing a totally different process.)

Eclipsing Variable Stars

Eclipsing variables are systems of two or more stars orbiting one another in which one star periodically eclipses (passes in front of) another. When this happens, the first star's light is partly or completely blocked. The light coming from the entire star system thus increases and decreases on a regular basis. Eclipsing variables are often subdivided into Algol variables, W Serpentis variables, and W Ursae Majoris stars, each named after the most famous star of its type.

Probably the best known of all eclipsing variables is the star

Algol, the second brightest star in the constellation Perseus. Algol (the name is an Arabic one meaning "the Demon") has been known for many centuries to vary regularly in brightness. In 1889 astronomers using a spectroscope confirmed that Algol is actually at least two stars, now called Algol A and Algol B. The two revolve around each other once every 68.8 hours. Algol A is the larger and brighter of the pair. From our line of sight, the two almost completely eclipse each other. When Algol B covers Algol A, there is a steep drop in magnitude; when Algol A passes in front of Algol B, there is another drop in brightness, but not as much as in the other case. It is now known that there is at least a third star in the system, Algol C, orbiting Algol A once every 1.86 years.

Pulsating Variable Stars

Pulsating variable stars periodically increase and decrease in brightness as their surface layers expand and contract. It is almost as if the star were breathing in and out. The period of the brightness variation is equal to the period of pulsation and is usually constant. There is often a change in the spectrum of the star, as the chemical composition of its surface layers changes with the pulsations. The luminosity of the star is directly related to its period of brightness.

There are many types of pulsating variables, including RR Lyrae stars, Cepheid variables, Mira variables, Delta Scuti stars, and irregular variables. The most famous of all variable stars, however, are the Cepheid variables. These are extremely bright yellow giant and supergiant stars with periods of from one to seventy days. Over seven hundred have been found in our Galaxy, and many thousands in the other galaxies of our Local Group.

Two general types of Cepheids are recognized by astronomers—classical Cepheids or Type I Cepheids, and Type II Cepheids. Classical Cepheids are younger and about 1.5 magnitudes brighter than Type II Cepheids. Classical Cepheids tend to have luminosity periods of five to ten days; Type II Cepheids have periods of twelve to thirty days.

The prototype classic Cepheid variable is the star Delta Cephei, discovered in 1784. Astronomers first noticed in the

1890s that its changes in brightness were accompanied by regular changes in its temperature and radius. Between 1908 and 1912 the astronomer Henrietta Leavitt discovered the relationship between the light variations and the period of pulsation for Cepheids.

When the absolute brightness of Cepheids was later determined, the astronomer Harlow Shapley realized that Cepheids could be used as interstellar yardsticks. Cepheids have since become invaluable indicators of distance to the nearest galaxies, such as the Magellanic Clouds and the Andromeda Galaxy. Cepheids and other variable stars, including RR Lyrae stars, are also important distance indicators within our own Galaxy.

The Birth of Stars

As we saw earlier, Walter Baade was able to use H-R diagrams to discover the existence of two different stellar populations with different ages. It is not surprising, therefore, that the evolution of different kinds of stars from birth through death can also be plotted on an H-R diagram. The H-R diagram "lifeline" of a star like the sun is quite different from that of a young A0 star like Vega (about twenty-four light-years from Earth) or a dim red M0 subdwarf like Kapteyn's Star (thirteen light-years distant).

The formation of solar-size stars is important for us personally—we're here because the sun is here. But the formation and evolution of more massive stars also is important for our existence. Their birth, rapid evolution, and violent death created the stuff of which we are made. So let's take a look at how both kinds of stars evolve and eventually die.

Stars are born in vast clouds of gas and dust called giant molecular clouds (GMCs). These clouds are made mostly of hydrogen gas, along with a smattering of helium and heavier elements like carbon, oxygen, silicon, and perhaps even iron. Some of the heavier elements are in the form of molecules, and others are tiny particles of dust. Something—the shock wave of a nearby exploding star, for example, or a spiral arm density wave passing through—causes part of the GMC to begin collapsing. Gravity begins to take control. The cloud fragment continues to contract, and its internal temperature begins to rise.

Within a few tens of millions of years, one of two things will happen. Either the temperature or density of the cloud will get high enough to counterbalance the gravitational pull and the contraction stops, or else the cloud's internal temperature will reach about ten million kelvins. If the former happens, the cloud fragment will become either a hot cloud or a brown dwarf. A brown dwarf is an almost-star, an object that's larger than a planet but not hot enough to be a star. But if the latter happens, the cloud *will* give birth to a new star.

Let's take a close look at one GMC fragment as it becomes a star. The GMC in question is about 120 light-years in diameter and contains about half a million solar masses of dust and molecular hydrogen gas. One segment of the GMC about two-thirds of a light-year wide begins to collapse under the impact of a shock wave from a nearby exploding star. This part of the GMC is often called a **dark cloud**, for the amount of gas and dust is so great that it blocks the light coming from more distant stars. As it collapses, its own gravity begins to take over. The cloud fragment continues to contract in size and increase in internal density and temperature. The density of the matter at the center increases faster than at the edge, so the cloud's center begins collapsing faster than its edges. The outer envelope is contracting, but more slowly than the center. At this point, with nothing to slow its descent, the collapse is in free-fall. Matter is falling into the center like garbage from the Apollo moon lander thrown down onto the lunar surface. The cloud fragment is now considered a **protostar**.

Deep down in the protostar, gas molecules and dust particles collide and release energy in the form of infrared radiation—heat. For a while the heat escapes freely into space, the cloud stays relatively cool, and the collapse continues unabated. The average temperature of the dark cloud fragment is about 10 K. Eventually, though, the density of matter at the center gets so great that the cloud becomes opaque. The infrared radiation cannot escape but is absorbed by nearby matter. The core begins to heat up, reaching a temperature of a few hundred kelvins. The pressure increases, and the core collapse slows down—but it doesn't stop.

Matter from the outer envelope continues falling in, heating the center of the collapsing protostar still more and increasing its mass and density. When the core temperature reaches 2,000 K, the hydrogen gas molecules break apart into individual atoms and soak up heat. The protostar's contraction speeds up as gravity again takes control, and its collapse again goes into free-fall. When all the hydrogen molecules are broken apart, the internal heat and pressure rise again, and the collapse slows down and finally stops. About a million years have passed since the collapse began.

The protostar continues to contract, but slowly. It is now twice the size of the sun and several times as bright, but no one can see it. The envelope of gas and dust surrounding the proto-star is still there, falling in and orbiting about the object but blocking it from anyone's view. However, the dust is still radiat-ing heat in the infrared part of the spectrum. Astronomers can see the protostar as an intense source of infrared radiation in the midst of the surrounding envelope and nearby GMC.

Like nearly every other object in the universe, the original cloud fragment probably had some small amount of internal rotation or spin. As it begins to collapse, that spin slowly starts to increase in order to preserve the cloud's angular momentum. This is a property of any rotating or spinning body. As the size of a rotating body gets smaller, it must spin faster in order to conserve this property of velocity and mass distribution. It is the same reason that a spinning ice skater spins faster when she pulls her arms in toward her body. In the case of the protostar and its outer envelope, another process also takes place. The central regions tend to collapse into a spinning spherical object. However, the outer regions, which include the surrounding envelope, begin to flatten out into a spinning pancake of dust and gas.

During all this time the protostar is getting its energy from gravitational energy. It is larger than it will be as a full-fledged star, and cooler, but also more luminous. This is because its surface area is larger and can radiate more energy than it will as a star. Convection is carrying the energy from gravitational contraction outward. The protostar at this stage is like a gargan-tuan bubbling ball of gas.

Eventually much of the envelope falls onto the surface of

the protostar. Its internal density and temperature continue to climb as it contracts. Its luminosity decreases as its surface area shrinks. Deep within the core of the protostar, the electrons are stripped from the nuclei of hydrogen, helium, and other elements. Naked helium nuclei are two protons and two neutrons bound together by a fundamental force of nature called the **strong nuclear force**. Hydrogen nuclei are nothing more than individual protons. In the intense heat of the protostar's core, the hydrogen nuclei fly about and bang into one another. Their mutually positive electrical charges keep them from touching, since like charges repel.

But when the temperature at the core reaches about ten million kelvins, the electrical forces are no longer sufficient to keep the protons apart. They smash into one another so violently, and are crowded together so closely, that the mutual electrical repulsion is overcome. Protons begin to "touch" one another. The strong nuclear force takes over, and protons begin to fuse together in pairs. A pair of fused protons has slightly less mass than two individual protons, for in the process of fusion some of the matter is turned into an enormous amount of energy. This is the process of nuclear fusion, which makes hydrogen bombs go boom and stars light up. The protostar lights up. It is now a star.

When the new star lights up, its luminosity and surface activity can vary widely and erratically. It will often begin blowing off layers of its surface in the form of a stellar "wind" of gas moving at speeds of up to several hundred kilometers per second. Whatever unconsolidated dust and gas still surround the new star as an envelope are swept away. These new stars, with masses from 0.2 to 2.0 times that of the sun, are **T Tauri stars** as they pass through this phase of their lives. The stellar wind is called a **T Tauri wind**. Some T Tauri stars lose as much as one ten-millionth of their total mass per year to the T Tauri wind. Sometimes the T Tauri wind is in the form of two collimated beams blowing away from the star's north and south polar regions. This is called a bipolar wind.

Not all the dust and gas in the new star's flattened disklike envelope get blown away by the T Tauri wind. Some of the matter in the envelope begins to clump together. Collisions bring dust particles together, and gravity begins keeping some together in larger and larger objects. The larger they get, the more of the surrounding gas and dust they sweep up. Sometimes a new star's T Tauri wind will kick in and sweep away most of a star's envelope before it can begin collecting into very large objects. And sometimes the process is well under way before the T Tauri wind begins. In the case of our own sun, some small percentage of its envelope had gathered into a sizable collection of both large and small objects before the T Tauri wind swept the remaining gas and dust from the scene. What was left behind, orbiting the sun, was a retinue of objects that eventually evolved into today's solar system.

The formation of all stars basically follows this scenario. However, there are some important differences for very large protostars, which evolve into giant and supergiant stars. For example, fusion reactions begin before the accretion of the envelope ends. Thus, the protostar stage is shorter for supermassive stars than for more normal-sized ones. When the fusion reaction begins, ultraviolet radiation emitted by the massive new star ionizes the hydrogen gas in the remaining cloud, creating what is called an **H_{II} region**. Only about half of the original cloud material falls into the central protostar. The rest gets blown away by the new massive star's wind and by the expanding H_{II} region.

The Death of Stars

In 1844 astronomer F. W. Bessel predicted the existence of a tiny but massive companion to Sirius. Bessel based his prediction on Sirius's unusual motions in the sky. Over many years of observations it had become clear that Sirius had a wobble to its movement. Bessel concluded that Sirius must be a binary star and that its "wobble" was actually caused by the fact that it and an invisible stellar companion were orbiting one another. To cause the wobble in Sirius's apparent path through the sky, that companion—though tiny—must have a mass similar to that of the sun. Eighteen years later astronomer Alvan Clark finally discovered a faint point of light in a telescopic image of Sirius. It was the Companion, now known as Sirius B.

Astronomers by then knew of the existence of tiny but extremely massive stars. However, there was still no good explanation for these objects, how they could form, and how small they could get. The answers to those questions finally came in the 1930s, particularly through the work of an Indian physicist named Subrahmanyan Chandrasekhar. Chandrasekhar explored what happens to matter in the interior of a star under the pressure of gravity. As a star contracts at the end of its life, the density at the center gets greater and greater. Electrons get stripped from the nuclei of atoms, and the nuclei and electrons exist together in an extremely dense "soup." As the density increases, the number of electrons per unit volume also increases. Soon the electrons are so tightly squashed together that they can exert a considerable pressure of their own. Physicists call this **degeneracy pressure**, and this state of matter is called **degenerate matter**. Degeneracy pressure is not like ordinary pressure, which depends on the temperature of the matter. Rather, it is dependent only on the density. Chandrasekhar discovered that the pressure exerted by electrons in such a gas can resist the force of gravity as long as the star has a mass no greater than 1.44 times that of the sun. This is called the **Chandrasekhar limit**.

A white dwarf is an extremely dense star caused when a large, massive red giant star collapses in on itself as it nears the end of its life. There may be as many as ten billion of them in our Galaxy alone. White dwarfs are very small and dense. They can have a density of around a billion kilograms per cubic meter—a tablespoon of white dwarf–stuff would weigh hundreds of tons on Earth—and have extremely strong gravitational fields. The radius of a white dwarf can be no more than that of the Earth. Sirius B has the mass of the sun, but its radius is only twice that of Earth. Some white dwarf stars, also known as ZZ Ceti stars, are pulsating variable stars. Their pulsation periods range from two to twenty minutes.

Novae (not to be confused with supernovae) are close binary stars in which one of the two stars is a white dwarf.

Matter from the second star is pulled onto the surface of the white dwarf by the white dwarf's gravitational pull. When enough matter has accumulated, it explodes in a huge fusion reaction and causes the white dwarf's light output to jump. This is the cause of the periodic appearance of a "new star" in the sky, called a nova.

One of the most astonishing secrets of white dwarfs, however, can never be directly seen. The star's immense gravitational field causes the atomic nuclei at the center of a white dwarf star to turn into a single mass embedded in the degenerate electron gas. If all that is left in the core of a collapsing and dying red giant star is carbon nuclei, then the resulting white dwarf star will be a mass of carbon nuclei. One is tempted to imagine a huge carbon crystal—a diamond, in other words—lying at the center of a white dwarf star. Of course, that's not literally true. There are no chemical interactions between the carbon nuclei and therefore no diamonds. But the image *is* intriguing.

What happens to a star that has a mass greater than 1.4 suns? The answer to that question was also worked out theoretically in the 1930s, mainly by Chandrasekhar, the Russian physicist Lev Landau, and the Bulgarian-American astronomer Fritz Zwicky. Electron pressure is not strong enough to hold back the crush of gravity. The dying star collapses still more. Soon the pressure at the center becomes so great that the electrons and protons are forced together and merge into neutrons and neutrinos. The neutrinos escape the star because they have no mass (or at least very little) and the star is practically transparent to them. This takes place at densities of around ten trillion kilograms per cubic meter.

When the density reaches one quadrillion (10^{15}) kilograms per cubic meter, the neutrons begin to "drip off" the atomic nuclei and form a separate "neutron gas" at the star's center. Finally, the squeeze of gravity causes the pressure at the center of the collapsing star to reach a hundred quadrillion (10^{17}) kilograms per cubic meter. The remaining atomic nuclei break apart into a gas that is about 80 percent neutrons, 10 percent electrons, and 10 percent protons. This "degenerate neutron gas" has enough gas pressure to halt the inward crush of gravity.

What has formed is a **neutron star**, a star made mostly of neutrons. Depending on its mass, its radius will be no more than ten to twenty kilometers. A neutron star with a twelve-kilometer radius will have a surface gravity a hundred billion times that of Earth. The inner one-kilometer core of a neutron star fifteen kilometers in radius probably has properties still unknown to science. The next ten or eleven kilometers are a neutron gas so dense that it acts like a fluid made of neutrons. The outer three kilometers are probably a mixture of neutron fluid and neutron-rich atomic nuclei, arranged in some form of crystalline lattice. The surface of a neutron star, then, is probably something like the interior of a white dwarf star. A neutron star could have an atmosphere a few meters thick and made mostly of protons, electrons, and gaseous iron.

Neutron stars do exist. The first was found in the summer of 1967 by British astronomers Antony Hewish and Jocelyn Bell Burnell. Named PSR 1919+21, it was emitting radio pulses every 1.33730113 seconds. The signal was so regular that at first some researchers half-jokingly dubbed it LGM-1, for "Little Green Men." Could it be a signal from aliens? It turned out not to be. Hewish and Burnell soon found several other such objects, and concluded they were pulsating stars. They named these objects **pulsars**.

Their conclusion was incorrect. It turns out that pulsars are in fact spinning neutron stars. These neutron stars are emitting pulses of radiation caused by electrons moving in the star's intense magnetic field. The radiation is beamed out from the star along its magnetic poles. When the magnetic poles are offset from the star's rotational poles, the beam of energy is swept around like a searchlight.

A pulsar is simply a neutron star that we have detected by its beam of radiation. Pulsars have now been detected at optical, gamma-ray, and x-ray wavelengths. Well over 350 have been detected so far. Astronomers estimate that the Galaxy may contain more than a million pulsars, or neutron stars.

Neutron stars are most likely formed in the aftermath of a **supernova** explosion. This is the most dramatic ending to a star's life. Its progenitor is an old, supermassive star that is much more massive than the sun, say, several solar masses. As it nears the end of its life, fusion reactions in its interior have converted the elements in its core into iron. At that point fusion cannot continue. Iron is the stopping point, since more energy is used in fusing iron atoms into heavier ones than is released by the process. Enormous gouts of energy are carried out of the star by a gargantuan flood of neutrinos. The "fire" goes out. There is nothing to keep up the internal pressure that supports the star's outer layers against the fierce pull of its own gravity. The star literally collapses. This collapse takes place in mere seconds.

When the matter above reaches the core, one of two things will happen. If the star is not too terribly massive, the core will be crushed into a neutron star, and the rest of the matter will "bounce" up and outward in a vast explosion of matter and energy. The result is what astronomers call a **Type II super-nova**. These supernovae take place almost exclusively in the spiral arms of galaxies. The star's matter is ejected into interstellar space in an expanding spherical shell traveling at five thousands kilometers per second or more. Its brightness can reach a magnitude of −17 or so. The supernova becomes, for a few weeks, the single brightest object in the entire Galaxy.

What is left behind in this case is a rapidly spinning neutron star and the expanding shell of gas and dust called a **supernova remnant** (SNR). Many SNRs have now been spotted in our Galaxy. The second youngest known SNR is the Crab Nebula. This is the remains of a star in our Galaxy that exploded in A.D. 1054. It was seen by observers in China and by Native Americans in North America. It is a beautiful crablike expanse of gas in the constellation Taurus. At its center is the Crab Pulsar, NP 0532, which is visible in optical, radio, gamma-ray, and x-ray wavelengths. The Crab Nebula is the single most studied object in the Galaxy.

The newest known supernova in our immediate region is Supernova 1987A, which appeared in the Large Magellanic Cloud in February 1987. It is rapidly overtaking the Crab as the most intensively studied single object in the Galaxy. Supernova

1987A is giving astronomers new insights into the origin of supernovae, and thus a deep look at how some stars finally die.

However, as mentioned, the supernova and neutron star constitute just one of two possible outcomes of a supermassive stellar collapse. Another possibility is that the star might be so massive that not even the degenerate neutron gas within its core has enough gas pressure to stop the crushing effects of gravitational collapse. If that is the case, then a supernova explosion will not create a neutron star. Instead, the core of the now-exploded star will continue to collapse . . . and collapse . . . and collapse. The result will be a black hole: an object that literally disappears from the universe.

9
THE MILKY WAY GALAXY

Even after Copernicus and Galileo replaced the Earth with the sun as the center of the solar system, astronomers by and large believed that the sun was at the center of the *universe*. By the beginning of the twentieth century, it seemed clear that the sun was just one of an immense number of stars that were part of a vast system astronomers called the **Galaxy**. The Galaxy *was* the universe, in this view. All that could be seen in the night sky, with or without a telescope, was part of the Galaxy. It was all part of one system, and that system was all that is. Beyond the Galaxy was nothing. However, even that view was eventually shown to be incorrect.

We now know that our Galaxy, the Milky Way Galaxy, is not the entire cosmos. It is one of uncounted billions of galaxies in a universe whose size staggers the imagination. It now appears that the universe is at least eighteen billion light-years in observable radius. And this brings up the matter of cosmic yardsticks.

Measuring Cosmic Distances

Astronomers use several different units when they measure distances. When measuring distances close to home, such as in our own solar system, the **astronomical unit** (AU) is commonly used. An AU is the mean distance from the sun to the Earth, about 93 million miles or 149.6 million kilometers. The planet Pluto, for example, travels as far as 49.2 AU from the sun in its orbit. The Oort Cloud of comets may stretch out as far as 100,000 AU or nearly 15 trillion kilometers from the sun.

187

For more cosmic distances, astronomers make use of the **light-year** and the **parsec**. A light-year is a pretty straightforward unit of distance. It is the distance a ray of light travels in one Earth year, about 5.88 trillion miles or 9.46 trillion kilometers. A parsec, by contrast, is a bit peculiar. It is defined by the apparent movement of stars across the sky caused by the Earth's revolution around the sun. This apparent movement is called **parallax**, and a parsec is a parallax shift of one arc second. In more understandable terms, a parsec is equal to about 3.26 light-years.

UNITS OF COSMIC DISTANCE

1 AU	= 93,000,000 mi
	= 149,597,870 km
	= 499 light-seconds
1 light-year	= 5.88 trillion mi (or 5.88×10^{12} mi)
	= 9.46 trillion km (or 9.46×10^{12} km)
	= 0.35 parsecs
	= 63,240 AU
1 parsec	= 3.26 light-years
	= 1.916×10^{13} mi
	= 3.025×10^{13} km
	= 206,162 AU

For various reasons, most astronomers use the parsec in their technical and scientific papers. Few nonastronomers are familiar with this unit of measurement. On the other hand, the vast number of science fiction movies and books, and the less vast but still considerable number of popular astronomy books, have made most of us familiar with the term *light-year*. We may not always have a clear understanding of how huge a light-year really is, but at least we've heard of it.

And how long *is* a light-year? Think of it this way: a 1990 Pontiac Grand Prix is about 5.5 meters in length from bumper to bumper. It would take 182 of these cars placed end to end to reach one kilometer, 69,960,800 of them to stretch from the Earth to the moon, and *1.72 quadrillion* of them to measure out a distance of one light-year. Here's another way to imagine a light-year. Suppose we find a stretch of vacant highway and take that same Pontiac Grand Prix out for a spin. We can probably reach a speed of 200 kilometers per hour. At that speed it would take us 47 billion hours to travel a distance of one light-year. That's about *5.4 million years* to travel a distance that takes light only one year to traverse.

A Galaxy Among Galaxies

Our Galaxy is not an isolated individual. For one thing, it has several companion or satellite galaxies. The best known are the Small and Large Magellanic Clouds. They're named after the sixteenth-century Spanish explorer Ferdinand Magellan, who spotted them during his circumnavigation of the Earth beginning in 1519. The Large Magellanic Cloud (LMC) is visually located in the Southern Hemisphere constellation of Dorado ("The Swordfish"), the Small Magellanic Cloud (SMC) in the constellation Tucanae.

The Galaxy also belongs to a "family" of galaxies bound together by mutual gravitational attraction. The **Local Group**, as astronomers call it, has about forty known members. The major members of the Local Group include the Milky Way Galaxy and its satellite galaxies, the Andromeda Galaxy (also known as M31) and its satellite galaxies (M32, NGC 147, NGC 185, and NGC 205), and the Triangulum Spiral (M33).

Galaxies come in many different shapes and sizes. Some appear spherical and nearly featureless. Others look like immense pinwheels of stars and dust. Still others seem to have bars of light embedded in their centers. Some galaxies have jets of matter spewing out of them, others look like Mexican sombreros, and others like misshapen splashes of spilt milk. For the most part, however, astronomers tend to classify galaxies in one of three categories first proposed by the American astronomer Edwin Hubble in the 1920s. These categories are spirals, ellipticals, and irregulars, abbreviated as S, E, and Irr.

Spiral galaxies are flat, like two pie plates glued edge to edge, with a central bulge or nucleus. The most distinctive ones have two spiral arms that seem to wind out and away from and around the center; these lie in the plane of the flattened disk.

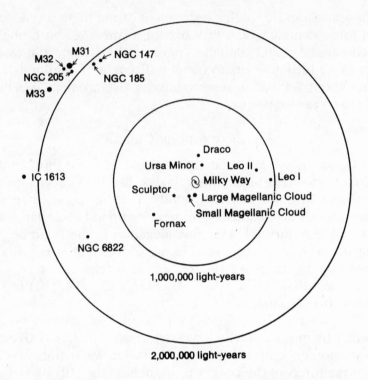

The Local Group of galaxies is the gravitationally bound "family" of galaxies to which our own Milky Way belongs. The Local Group includes our own Galaxy's satellite galaxies, the Large and Small Magellanic Clouds, several nearby dwarf galaxies, and M31, the famous Andromeda Galaxy.

Illustration from Robert T. Dixon, *Dynamic Astronomy*, fifth edition, 1989. Reprinted by permission of Prentice-Hall, Inc., Englewood Cliffs, New Jersey.

Some spiral arms appear to emerge from a bar centered on the nucleus. This has led to the creation of two subcategories for spirals, S and SB (for barred). These categories are further subdivided into Sa (and SBa) through Sc (and SBc) by the size of the nucleus and the tightness with which the arms are wound around the galactic structure. Sa and SBa galaxies have the largest nuclei and the most tightly wound and smoothly flowing arms; Sc and SBc have very small nuclei and fragmented arms.

Elliptical galaxies are just that: more or less elliptical or spherical balls of stars with no other distinctive structure. The only other clue to their structure is that their brightness or luminosity decreases the farther outward one moves from their center. The rate at which the brightness decreases follows a simple mathematical formula first developed by Hubble. Elliptical galaxies are broken into subclasses designated by a number from 0 to 7, which follows the *E*. The larger the number, the less spherical the elliptical galaxy. An E0 galaxy would be perfectly smooth and spherical, while an E7 elliptical would be highly distorted from a perfect spherical shape. Elliptical galaxies appear to be governed by little more than gravity. The orbits of their stars may be elliptical, but they're smooth. The galaxy as a whole usually has only a small overall rotation. Little new star formation is going on anywhere.

Irregular galaxies include everything that isn't a spiral or an elliptical galaxy. They tend to look very patchy and ofttimes amorphous. Some contain distinctive bar structures or patches that look like fragments of spiral arms. Irregular galaxies made mostly of blue stars were originally classified as Irr I, and those with mostly red stars as Irr II. Irr II galaxies are now known to be mostly galaxies that are highly perturbed either by collisions with another galaxy (**interacting galaxies**) or by a violent internal burst of star formation (**starburst galaxies**).

The Milky Way Galaxy is a spiral galaxy of type Sb or Sc; the classification is still open to debate. It is fairly typical of either classification. Its visible spiral disk structure is about 130,000 light-years in diameter from edge to edge, and about 9,800 light-years from north to south through the thickest part of its nucleus.

However, there is considerably more to a galaxy's structure than its visible shape. In fact, one of the most important parts of a galaxy is invisible. Since our journey will be to the center of *our* Galaxy, let's take a look at its overall structure before we begin our long dive inward.

The Galactic Structure

The overall structure of our Galaxy has four major parts: the **corona**, the **halo**, the **disk**, and the **central bulge**. The entire Galaxy rotates about an axis that passes through the center. The disk rotates fairly rapidly, and the halo somewhat more slowly, about this axis. At the sun's distance from the Galactic center, the systematic rotation of disk stars is about 220 to 250 kilometers per second. At that same distance a halo star is orbiting the center at a velocity of only about 50 kilometers per second.

The corona is the single largest part of the Galaxy, extending from about one hundred thousand light-years to as far as three hundred thousand light-years from the Galactic center. It is also the Galaxy's most mysterious component. Astronomers now calculate that as much as 90 percent or more of the Galaxy's mass must be "hidden" or "dark." By this, astronomers and physicists mean that it does not seem to interact with the rest of the matter and energy in the Galaxy in any way except through gravitation. We sense its presence through its gravitational pull on the stars in the outer Galaxy. The corona is the

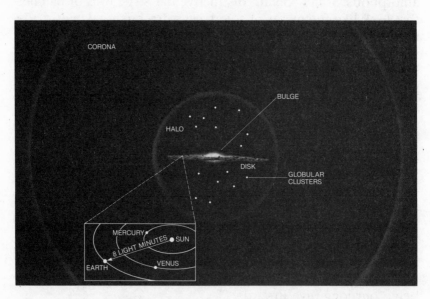

The parts of the Milky Way Galaxy include the mysterious Galactic corona, the Galactic halo with its family of globular clusters, the Galactic disk, and the central nuclear bulge. In the center of the bulge, at the Galactic core, is believed to exist a massive black hole.

Illustration from Jay Pasachoff, *Contemporary Astronomy*, third edition, Saunders College Publishing, a division of Holt, Rinehart and Winston, Inc., 1985. Reprinted by permission of the publisher.

location of nearly all this "missing mass," which includes more than ten times as much mass as all the visible stars, gas, and dust in the rest of the Milky Way. The corona may well include more than a trillion solar masses of dark matter. The Milky Way Galaxy may thus have a mass of more than two trillion suns.

The next major part of the Galaxy's structure is the Galactic halo. It extends about eighty thousand to a hundred thousand light-years out from the Galactic center. The halo is composed of more "conventional" matter—dust, gas, and stars. It also includes globular clusters, the spherically symmetrical and compact star clusters that contain anywhere from tens of thousands to several millions of stars. Globular clusters are part of the Galaxy's halo, but the orbits of many of them extend out to at least a hundred thousand light-years from the center. About 125 globular clusters are known. Nearly all the stars in the halo, including those in the globular clusters, are old metal-poor Population II stars formed at least twelve billion years ago. The Galactic halo as a whole probably has a mass of about 120 billion suns.

The bulk of the Galaxy's stars, interstellar gas, and dust is found in a thin disk that is most visually notable for its spiral arms. The disk lies more or less along the Galactic equatorial plane. Its radius is about 65,000 light-years from outer edge to the Galactic core. Its maximum thickness at the center is about 13,000 light-years. The vast majority of the stars in the disk are fairly young Population I stars, ranging from about three to six billion years old. About half of the matter in the Galactic disk is concentrated in **giant molecular clouds** (GMCs), which have masses of from one to ten million suns and life spans of a hundred million to a billion years.

The GMCs are distributed along the Galaxy's **spiral arms**. There is still some debate about how many spiral arms the Galaxy actually has. In fact, it is quite possible that the "arms" could more properly be called arm segments.

The fourth major part of the Galaxy is its central nuclear bulge. This is a flattened and spheroidal nucleus at the geometrical and rotational center of the Galaxy. The bulge is about 9,800 light-years thick along the north-south Galactic axis and about 16,000 light-years thick along the Galactic plane. The total mass of the nuclear bulge is estimated at about 140 billion suns.

At the center of the Galactic bulge is the Galactic core. It is about a thousand light-years in diameter. And at the very center of the core, believe many astronomers today, is a supermassive black hole with the mass of at least a million suns.

Now we begin our journey to the center of the Galaxy. The time is the future, a future in some ways like our present and in some ways very different. People still play tourist and take vacations, taking boxloads of pictures and videos of their travels and the strange sights they see. But where they travel on their vacation is not just Egypt or the Grand Canyon—or even the tired old sights of Tycho Crater and Valles Marineris or the sulfur volcanoes of Io System Park. Oh, no . . .

10
THE OUTER REACHES OF THE GALAXY

Our journey to the center of the Galaxy begins at a place and time so distant that it takes light from the core more than three hundred thousand years to reach us. Were this the twentieth century, of course, we would be very limited in our ability to travel about the universe. It would take us many millions of years to travel from the outer edges of the Galaxy to its central regions. In our day, however, we are no longer bound by the light-speed limitation.

Our metaship uses the energy that lies hidden in the very structure of the vacuum itself. By coupling one pole of its drives to the vacuum and its energy and the other pole to the supergravity flux of the universes, the metaship can "skip" across spacetime like an expertly thrown stone across the surface of a pond. The "pond" is this universe. The "stone" is the metaship. It skips (or, as the metaship refers to it, "Jumps") at many multiples of the speed of light. We who ride on the stone can flit about as we wish.

Our journey begins here, at the outermost edge of the Galaxy. We enter the corona.

In his Foundation series of science fiction novels, Isaac Asimov used the phrase *opposite ends of the galaxy* to great effect. The visible Galaxy is disk-shaped and thus does not have "opposite ends" in the literal sense. However, a disk *does* have an outer edge and a center. Those were the "opposite ends" that Asimov's character Hari Seldon cleverly referred to when he spoke of the locations of his two "Foundations." One of the Foundations in the novels was well known. It was the publication center for the *Encyclopedia Galactica* and the driving force behind the renaissance of Galactic civilization. The other Foundation, however, was shrouded in mystery.

The real Milky Way Galaxy doesn't actually have an "outer edge." The Galactic disk trails off, fading into the halo. And no one really knows for sure where the mysterious Galactic corona ends and interstellar space begins.

However, if we're willing to take some literary or imaginative license, à la Asimov, then it is in a way true that the two most mysterious parts of our real-life Galaxy are its two "opposite ends": the center of the Galaxy and its outermost realms. We begin our journey to one mystery—the center of the Galaxy—at the edge of another—the corona.

Quintessence and Dark Matter

Several of the ancient Greek philosophers believed that objects in the heavens like planets and stars were made of a fifth primordial element called quintessence. Everything else was made of some combination of earth, air, fire, and water. Galileo's telescopic discovery of mountains on the moon shook to its foundations the idea that heavenly objects were somehow different in composition from those of the Earth. Through the nineteenth century additional evidence accumulated that cosmic objects like stars are made of the same stuff as things on Earth. Fraunhofer's discovery in 1815 of absorption lines in the sun's spectrum was one piece of evidence. Kirchhoff's explanation in 1862 that individual chemical elements have unique spectral line "fingerprints" was another.

However, the idea of a "quintessence" never quite went away until the end of the nineteenth century. The discovery that light acts like a wave led James Clerk Maxwell to assert that all of space was permeated with an invisible substance he called the "ether." Maxwell could not believe that electromagnetic waves could travel through empty space, with no medium to support them. Ergo, he said, there must be an ether—a direct philosophical descendant of the Greek quintessence. Michelson and Morley demolished the ether hypothesis in 1887 with their series of experiments looking for increases and decreases in the speed of light. The final explanation of how light waves can travel in a vacuum did not come until Einstein's theory of special relativity in 1905.

By the 1930s the nonexistence of anything like a "quintessence" seemed assured. Everything in the universe, from distant stars and nebulas to grains of sand on the beach, are made of the same few dozen elements. And all objects, except for nonexistent blackbodies, radiate light of some kind. Then in 1932 came the first hints that our own Galaxy might contain something that is nonluminous and therefore not normal matter.

The first piece of evidence came from the work of a young Dutch astronomer named Jan Oort (1900–   ). Oort had already made a significant contribution to Galactic astronomy. In 1927 he had confirmed the earlier hypothesis of Bertil Lindblad that the Galaxy is rotating. He would later go on to use radioastronomy to map the Galaxy's spiral arms. In 1932, meanwhile, he was following up his work on the Galaxy's rotation.

Oort measured the positions and motions of a number of stars lying outside the visible disk of the Galaxy. Then he used that information to calculate how much mass must lie inside their orbits to produce their observed motions. This amount is called the **Oort limit**, and it is equal to about 0.03 of a solar mass per cubic light-year. Next Oort added up the masses of the visible stars in the Galactic disk. The result was surprising: the actual mass present in the Galaxy seemed to be 50 percent *less* than what was needed to cause the actual movements of the stars that lay outside the visible Galactic disk. Oort added a fudge factor to correct for the discrepancy, assuming the existence of many dim and uncounted stars in the Galaxy. However, other studies in the years since have only served to confirm Oort's initial discrepancy.

The second piece of evidence for the existence of some strange form of invisible matter in the universe came from Fritz Zwicky, the Bulgarian-American astronomer who with Walter Baade would later prove the existence of supernovae. In 1932 Zwicky was busy measuring the positions and movements of galaxies found in large clusters. In effect, he was doing the same thing that Oort was doing, but on a much larger scale. The results, however, were the same. Zwicky found that galaxies in clusters were orbiting around each other at velocities far greater than could be explained by the matter visible in the clusters. The discrepancy was greater by a factor of a hundred or more, much larger than even the disparity found by Oort.

The Corona Proposal

Zwicky's figures were rejected by many astronomers, who suspected that his calculations were somehow not quite accurate. Nevertheless, other observers seemed to come up with similar results. At the same time, the mass discrepancy in our own Galaxy discovered by Jan Oort persisted. By the early 1970s it had become clear that the visible gas clouds, dust clouds, stars, and any reasonable number of faint red dwarf stars simply do not add up to more than 10 percent of the Oort limit. Ninety percent or more of the Galaxy appears to consist of matter that is invisible on all electromagnetic wavelengths. Only its gravitational pull gives it away. It is "dark matter," and it constitutes the "missing mass" of the Galaxy—and indeed of most of the universe.

One question that astronomers asked was, Where is it? The now-accepted answer was first proposed in 1973. That year astronomers Jeremiah Ostriker, P. J. E. Peebles, and Amos Yahil of Princeton and Jan Einasto of Estonia first suggested the existence of an outer part of the Galaxy they called the **corona**. The four researchers noted what Oort had first discovered back in 1932. Stars, dust, and gas clouds in the outer regions of the Galaxy are traveling along their orbits around the center much faster than they should, given the Galaxy's commonly assumed mass. This is strikingly true of globular clusters having orbits extending out to at least 196,000 light-years from the Galactic center. The most distant of these are traveling at radial velocities much greater than would be expected if they were in empty intergalactic space. In fact, their observed velocities would carry them *completely out of the Galaxy* without the presence of some invisible mass. However, they *are* clearly gravitationally bound to the Galaxy. Their high velocities imply a minimum Galactic mass of six hundred billion solar masses, much greater than an estimate based on the visible number of stars, dust, and gas. Clearly, the four researchers said, an outer Galactic corona of "dark matter" exists. It has to.

Most astronomers now accept the existence of a Galactic corona. It is the single largest part of the Galaxy. It is also the most mysterious and the most recent to be identified. The corona is the location of nearly all the Galaxy's "missing mass." Some researchers at first suggested that the corona is a vast sphere of dark matter that extends to at least 196,000 light-years from the center of the Galaxy.

This is not much farther out from the center than the Galactic halo. There is evidence, though, that the corona may extend much farther. The velocities and paths of nearby dwarf galaxies, as well as the Milky Way's nearby satellite galaxies, the Magellanic Clouds, can be explained only if the corona has a radius of around 326,000 light-years. Moving around and through this extended corona of mostly dark matter are objects made of "normal matter." They include several globular clusters, the two Magellanic Clouds, and at least seven other dwarf galaxies that are gravitationally bound to the Milky Way.

Another Piece of Evidence

Direct evidence for the existence of the Galaxy's corona was found in 1983 by M. R. S. Hawkins, an astronomer at the Royal Observatory in Edinburgh, Scotland. Hawkins's evidence came from spectroscopic observations of a star named R15. This star is an **RR Lyrae star**, and was discovered during a search for faint variable stars. As we saw earlier, variable stars are those which vary in brightness. Some have regular periods, while others are irregular in their brightness variations.

RR LYRAE STARS

RR Lyrae variables were first discovered in 1895 by Solon Bailey. The entire class is named for the star RR Lyrae (found in the constellation Lyra, "the Lyre"), which was discovered in 1899. RR Lyrae stars are pulsating variables—their change in brightness is caused by a regular pulsation in the size (and therefore the luminosity) of the star. They are very old giant stars, belonging to the Population II class, and are mainly found in globular clusters. They usually have periods of less than one day. Their spectral types range from A7 to F5, and they have an absolute magnitude of around +0.5. By comparison, the sun is spectral type G2 and has an absolute magnitude of only +4.2. So RR Lyrae variable stars are hotter, bluer, and brighter than the sun. Because of their strong absolute brightness, RR Lyrae stars can be used as cosmic yardsticks for up to 600,000 light-years.

R15 lies about two hundred thousand light-years from the center of the Galaxy and about one hundred fifty thousand light-years above the Galactic plane. It is not part of a globular cluster but is an isolated star. Hawkins's discovery about R15 was fairly simple in nature but powerful in its consequences. He found that the absorption lines in the star's spectrum showed a Doppler shift. R15 is moving toward Earth at a velocity of about 465 km/sec (more than a million miles per hour). This velocity could be translated into a lower limit for the mass of the Galaxy lying inside R15's orbit around the Galactic center. The answer is astonishing: if R15 is in fact gravitationally bound to the Milky Way, then our Galaxy must have a mass of at least *1.4 trillion* suns. That's ten times greater than the estimates of the Galaxy's mass commonly accepted for the last fifty years.

In fact, the actual mass of the Galaxy could well be higher than that. There is currently no way to tell whether R15 has a large velocity at right angles to our line of sight. If it does, the value for the Galaxy's mass will be higher than 1.4 trillion solar masses.

Dark Matter: Dwarfs, Jupiters, or Neutrinos?

Clearly, the corona is not only huge in physical extent, it is huge in mass as well. But what is it?

Speculations about the nature of the dark matter in the corona (or the Galaxy and, for that matter, the rest of the universe) can be divided into two categories: "reasonable" and "far out." The "reasonable" category includes white dwarfs, black dwarfs, brown dwarfs, and "jupiters." The "far out" category includes black holes of various sizes and various theoretical subatomic particles that have odd-sounding names and have not yet been discovered.

White dwarfs, as we've already discussed, are the dying cores of a generation of fairly massive stars, some of which first formed many billions of years before the birth of our own sun. Although they have the mass of our sun, they are very small and faint, and therefore are almost impossible to detect telescopically. Black dwarfs are white dwarf stars that have finally burned out. When all the energy has been squeezed out of a white dwarf by its gravitational contraction and all its radiation has leaked out into space, the white dwarf finally goes "black." It is not a black hole, though; it is simply the lightless husk of a star that has finally and irrevocably died.

Unlike white dwarfs and black dwarfs, brown dwarfs and "jupiters" are not the remains of stars at all. They are substellar objects, spherical concentrations of matter that never got quite large enough for fusion reactions to begin in their cores. Whatever faint light or energy they emit is caused by gravitational contraction and heat.

Brown dwarfs are defined as substellar objects with masses less than 0.05 to 0.08 solar masses. The upper limit of 0.05 solar masses is the amount of mass needed for thermonuclear reactions to trigger off in the center of a stellar object. The faintest known red dwarf stars would have masses in this range. Objects with a mass of less than 0.08 suns will shine for only about a hundred million years, with their energy coming only from gravitational contraction. In the late 1980s several astronomers found evidence for the presence of brown dwarf objects circling several nearby stars. These objects have not yet been imaged; it was the wobbling motion of the stars that gave them away. (Other researchers, including astronomer Ben Zuckerman, may recently have detected their infrared radiation.) And these brown dwarfs, it should be noted, were not isolated objects in space. They were companions of more normal-sized stars.

Some astronomers theorize the existence of substellar objects even smaller than brown dwarfs. We could call these objects "jupiters." Jupiter is the largest planet in our solar system. Its mass is about 1.9×10^{27} kg, about one-thousandth of a solar mass. (The sun's mass is 1.99×10^{30} kg.) A "jupiter," therefore, would be a substellar object Jupiter's size or larger but considerably smaller than a brown dwarf. In other words, such objects would be very big planets.

Two significant problems exist with these various "reasonable" explanations for dark matter. The first has to do with numbers. To account for all the missing mass in the Galaxy, including the corona, would require a hundred billion white or black dwarfs, or ten to a hundred *trillion* brown-dwarfs or jupiter objects. While most astronomers assume some of these objects are certainly floating around in space, few astronomers believe there are hundreds of trillions. There is simply no convincing evidence for the existence of that many white, black, or brown dwarfs.

The second problem is the lack of evidence for the simple existence of some of these conventional objects. White dwarfs clearly exist, of course. Black dwarfs probably do too. The universe is now believed to be eighteen to twenty billion years old. According to astronomer Mark Morris, it should take "only" a hundred million to a billion years for a white dwarf to completely cool off and become a black dwarf. It's just that they would be extraordinarily difficult to detect. Some intriguing evidence has appeared in the last several years for the existence of brown-dwarf objects circling nearby stars. However, there is no convincing evidence for the existence of any isolated brown dwarfs in the Galaxy. Finally, no evidence exists at all for the existence of black dwarfs or free-ranging, unattached "jupiters" anywhere.

One other possible "reasonable" explanation for dark matter is the neutrino. This is a subatomic particle whose existence was first proposed in 1930 by Wolfgang Pauli. He postulated its existence in order to explain the apparent disappearance of energy during the process of beta decay. An isolated neutron is not a stable particle. It eventually decays into a proton and an electron, but it seems that some energy is left over, or not accounted for. Pauli suggested that the leftover energy is carried off by a previously undiscovered particle, which he dubbed the "neutrino."

Neutrinos were finally detected in 1956. Since then physicists have detected or postulated the existence of three types of neutrinos: electron-neutrinos, muon-neutrinos, and tau-neutrinos. Electron-neutrinos are the most common and are what physicists are usually referring to when they speak of "neutrinos."

All three kinds of neutrinos, like photons, have been assumed to have no electrical charge and no mass. In recent years, however, some scientists have begun to suspect that neutrinos *do* have mass. It may not be much—perhaps it could be measured in energy terms as only a few electron volts of mass—but at least some.

Massive neutrinos would explain some puzzling problems with certain theories of physics. They could also help solve the problem of the universe's missing mass. If neutrinos have mass, then they can be affected by gravity. And that means the Galaxy's corona could be largely composed of massive neutrinos.

However, the various experiments to detect any mass for neutrinos have been inconclusive. Some recent experiments by Russian physicists have suggested that neutrinos have a mass of a few thousandths of an electron volt. Others have found no such thing. The question is still up in the air—along with the question of the nature of the dark matter in the corona.

The Corona's Dark Matter: Black Holes and Axions?

It appears that the "reasonable" explanations for the nature of the dark matter are flimsy at best. That leaves the "far-out" proposals. They include black holes and bizarre subatomic particles with equally strange-sounding names.

Some astronomers have suggested that much of the missing mass in the Galaxy (and the universe) could be in the form of black holes. We know that black holes can theoretically come in many different masses, from asteroid or Earth mass, to stellar mass, to masses equaling millions of suns. Black holes with stellar masses will not do, since they would reveal their presence by the energy emitted by the gases falling into them. A supermassive black hole at the Galactic center would account for several million suns' worth of missing mass in our Galaxy, but that's still a million times less than the total amount of dark matter thought to exist. A million such black holes scattered throughout the corona would long ago have revealed their presence by the massive gravitational perturbations they would cause to the orbits of the globular clusters.

That leaves mini–black holes. These theoretical objects would have masses like asteroids or planets. They would be very tiny, with "radii" no more than a centimeter or so. Thus, they would not give themselves away either by their individual gravitational pull or by any radiation from in-falling gas. The one big problem with mini–black holes is the same as with jupiter objects. There is simply no convincing evidence that they even exist.

This leaves us with exotic subatomic particles. And they are pretty far out indeed. They have to do with the attempts by physicists to construct a "unified theory" of the universe.

Such a theory would explain all four known forces of the universe as aspects of one original, primeval force, which existed in the first few instants of the Big Bang at the beginning of space-time. As the universe expanded, the primeval force "broke" into the four forces we know today: weak nuclear, strong nuclear, electromagnetic, and gravitational. Each force has a particle that "carries" it from object to object, much as a letter carrier delivers mail from one person to another. The photon, for example, is the particle that "carries" the electromagnetic force. The gluon carries the strong force, which binds quarks together into protons, neutrons, and several other subatomic particles. The W and Z particles are the carriers of the weak force, which governs certain forms of radioactive decay. The graviton is the carrier particle for the gravitational force. Photons and the W and Z particles have all been detected. Gluons and gravitons have not, but their existence is assumed by nearly all physicists today.

Einstein spent the latter part of his life searching for a unified field theory, but failed. Other physicists in more recent years have been somewhat more successful, but the ultimate unified theory still eludes them. Several highly technical theories, which are partial attempts at a unified theory, require the existence of unusual counterparts to the better-known particles like photons and gravitons. These weird particles have names like **photino** and **gravitino**, because they have certain kinds of relationships to the more normal photon and graviton.

Another such theoretical particle is called an **axion**. An axion is a particle that would make the theory of the strong nuclear force (called quantum chromodynamics, or QCD) consistent with the results from actual experiments. An axion would be a subatomic particle that has very little mass and rarely interacts with normal matter.

Still other bizarre (and theoretical) particles have been grouped together under the general acronym of WIMPs, which stands for "weakly interacting massive particles." Because they are "weakly interacting," they would not affect the normal nuclear reactions of protons and neutrons. Thus, WIMPs would not give away their presence by interfering with fusion reactions in stars. They also would not interact with electromagnetic fields, so they would not radiate photons. WIMPs cannot be "seen." Unlike axions, however, WIMPs are massive particles. The presence of enough WIMPs in the corona would have a significant gravitational effect on objects made of normal matter. WIMPs, photinos, gravitinos, and axions are considered by some physicists and astronomers to be likely candidates for the dark matter of the corona.

There is, however, one very important problem with this explanation. These are all particles whose existence is *predicted* by one or more theories. But no one has actually detected the existence of WIMPs, photinos, gravitinos, or axions.

The corona does exist. It gives itself away by its gravitational pull. It is made of something we call dark matter. But the nature of dark matter remains a mystery for future astronomers or physicists to unravel.

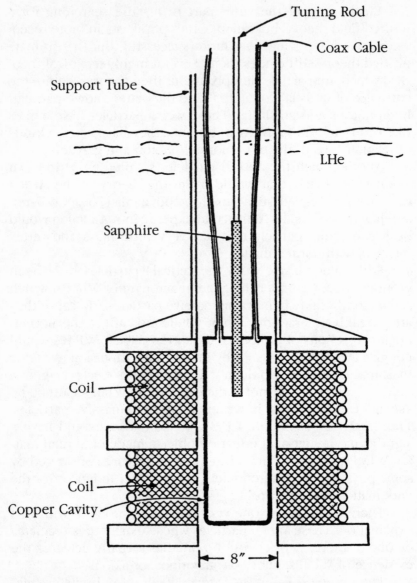

Is the Galactic corona composed of axions? These still-hypothetical subatomic particles may be the "dark matter" that comprises more than 90 percent of the mass of the Galaxy—and indeed of the entire universe. Bruce Moskowitz and his colleagues at Brookhaven National Laboratory have looked for axions using a Galactic axion detector, shown here in a schematic drawing.

Illustration courtesy of Bruce Moskowitz, Brookhaven National Laboratory.

As we travel some two hundred thousand light-years in from the outer edges of the corona, we seem to be traveling through a black void. Our highly sensitive telescopes and energy sensors detect the occasional presence of a stray star or a black hole and faint white and black dwarfs scattered about the vast volume of space. Here and there we see a spherical globular cluster whose eccentric orbit has swung it far from the Galactic center and out into near-intergalactic space. Mostly, though, space appears empty. It is only our gravity detectors that espy the presence of dark matter, which envelopes the distant visible Galaxy like a vast sea swirling around a tiny island. The corona is that mysterious, invisible sea.

Some of the more "sophisticated" passengers chuckle at the wild attempts of twentieth-century astronomers to explain the nature of the dark matter in the corona. Today the answer seems so . . . obvious. But then, others gently point out, what was also self-evident to our twentieth-century ancestors—such as supermassive black holes and self-aware, artificially intelligent computers—would have been a total mystery to the Ancients. It's all relative.

We turn our attention inward. The island we call the Milky Way, our Galaxy, lies dead ahead.

11
THE GALACTIC HALO

Our metaship takes us deeper and deeper into the Galaxy. Because we lie at the very edge of the Galaxy's gravity well, at first we move slowly. Our translight drive can initially take us only on relatively small Jumps of ten thousand light-years or so, and the metaship must rest for several standard days between each Jump. Unlike the ancient kangaroo, she must "catch her breath" between leaps. She is also charging her internal power sources with excess energy. We will need it all to shield us from the enormous forces that lie waiting at the core.

As we move further into the gravity well and are surrounded by more and more gravitationally alive matter, the metaship is able to use more of the energy that lies squeezed into the interstices of the vacuum. She takes larger and larger Jumps. Still and all, it takes us fifteen Jumps and nearly two complete lunations before we arrive at the inner edge of the corona.

There is nothing here to mark the transition, of course. No boundary line or wall of energy marks the point where the corona ends and the next part of the Galaxy begins. However, astronomers do have an imaginary boundary line. It is that place in space touched by the outer parts of the orbits of the globular clusters. Their apogalacticons mark the beginning of the region called the halo. At a hundred thousand light-years from the center, we enter the outer edges of the normal-matter Galaxy.

The corona shades into the **Galactic halo**, the next major part of the Galaxy's structure. It is also called the Galactic spheroid. Its outer boundary is marked by the outermost parts of

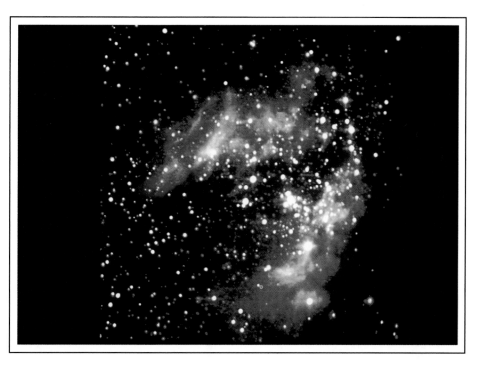

M17 is a massive region of star formation in the Galaxy, about 7,000 light-years from Earth. This false-color infrared image reveals some of the brighter details in M17. The total amount of energy radiated by the stars within M17 is some six million times greater than that of our sun.

National Optical Astronomy Observatories; photograph by Dr. Ian Gatley

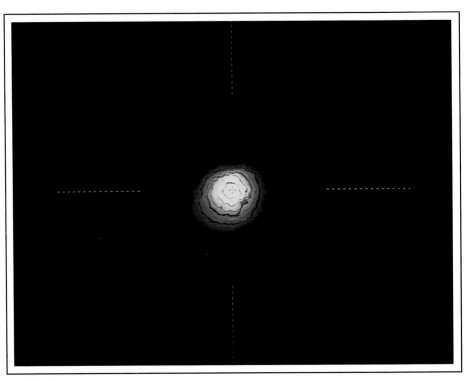

A halo of gas can be seen around the star Betelgeuse in this computer-enhanced image. Betelgeuse is a red supergiant in the constellation Orion, about 390 light-years from Earth. Its spectral type is M2 Iab, its diameter 800 times that of the sun. The shell of gas seen in this image is one of three puffed out by Betelgeuse over the last 100,000 years as it nears the end of its life. The most distant shell has a radius of more than four light-years. Betelgeuse is a prime candidate to be the next star in our Galaxy to explode as a supernova. It could happen 100,000 years from now—or tomorrow.
National Optical Astronomy Observatories

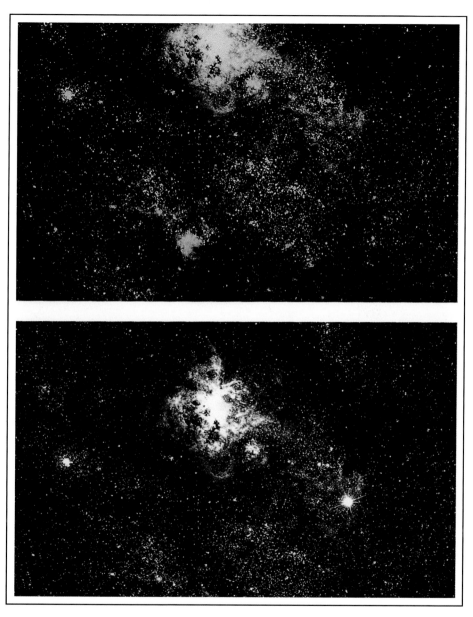

"Before" and "after" photographs of Supernova 1987A, which is located in the Large Magellanic Cloud (LMC). The top image was made in 1969 by Dr. Victor Blanco. The bottom image was taken on February 26, 1987, by Dr. Wendy Roberts shortly after the supernova was first discovered. The supernova can clearly be seen in the bottom image, to the right and slightly below the Tarantula Nebula. Supernova 1987A is the first supernova to be observed in or near our own Galaxy in almost three hundred years.

National Optical Astronomy Observatories

The Large Magellanic Cloud (LMC) is a satellite of our own Galaxy. It is visible with the naked eye from the Southern Hemisphere. Along with the Small Magellanic Cloud (SMC), it is named for the Portuguese explorer Ferdinand Magellan, who observed them in 1519. The LMC is a Type I irregular galaxy with a diameter of about 39,000 light-years. It lies in the constellation Dorado, some 163,000 light-years from Earth—likely within the Galactic corona. It has a total mass of about ten billion suns. The LMC contains many fairly young Population I stars and a larger proportion of gas than our own Galaxy. Because it is so close to us, astronomers have been able to study it in great detail. The large gaseous nebula is the Tarantula Nebula (30 Doradus).

National Optical Astronomy Observatories

The Small Magellanic Cloud (SMC) is another companion galaxy of our own. This dwarf irregular Type I galaxy is located in the constellation Tucana, and in the Southern Hemisphere it is visible to the naked eye. It lies about 195,000 light-years from Earth, contains about two billion solar masses, and is about 20,000 light-years in diameter. Observations made in 1983 have led some astronomers to suggest that the SMC we see is actually two galaxies superimposed along our line of sight. The second, sometimes called the Mini Magellanic Cloud, lies about 32,000 light-years farther away. Like the LMC, the SMC contains many Population I stars and a high proportion of gas and dust.

National Optical Astronomy Observatories

The Andromeda Galaxy (M31) is one of the most famous galaxies in the night sky. M31 is a member of the Local Group. It is about 110,000 light-years in diameter and contains over 300 billion stars. M31 is a Type Sb spiral with a bright, ellipse-shaped nucleus. It is about 2.2 million light-years from us. In the foreground is M32, one of the satellite galaxy companions of M31.

National Optical Astronomy Observatories

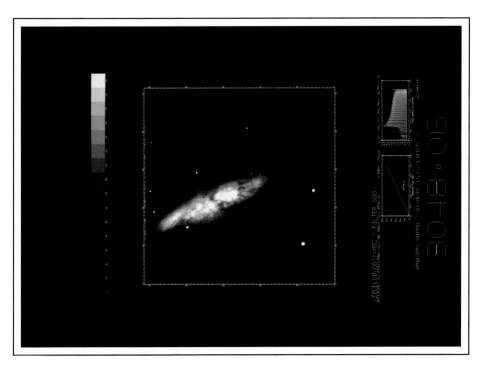

M82 is a Type II irregular galaxy in the constellation Ursa Major. The colors in this photograph, which is a combination of three images made with a charge-coupled device (CCD), approximate the true appearance of M82.
National Optical Astronomy Observatories

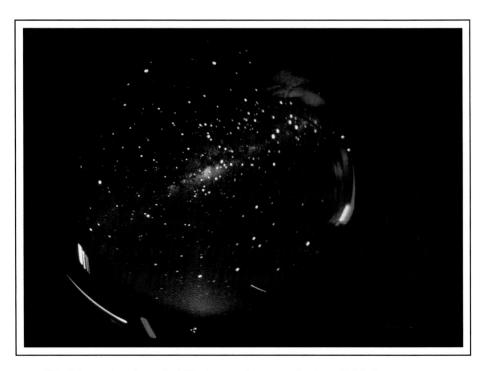

This fish-eye view shows the Milky Way arching over the Cerro Tololo Inter-American Observatory in Chile.
National Optical Astronomy Observatories

the orbits of the globular clusters. The farthest point of the orbit of a star or other object (like a globular cluster) from the Galactic center is called the *apogalacticon*. The nearest point of the orbit is called the *perigalacticon*.

The overall distribution of globular clusters forms a rough sphere that is centered on the Galactic center. They actually travel in highly elliptical orbits around the Galactic center. Their orbits take most of them as far as forty thousand to sixty thousand light-years from the center, although some globular clusters have apogalacticons that extend as far as a hundred thousand light-years out. As they move in their orbital paths, globular clusters dive through the Galaxy's disk. Over billions of years their repeated passages through the disk has swept them clear of all the dust and gas they may have once had. Globular clusters today are almost completely lacking any gas and dust clouds.

Nearly all the stars in the halo, including those in the globular clusters, are very old. These Population II stars were formed at least twelve billion years ago, when the Galaxy itself was still young and the cloud of gas from which it was born was still spherical. The most conspicuous of these stars, other than the globular clusters, are RR Lyrae variable stars. The Galactic halo as a whole probably has a mass of about 120 billion suns.

Besides globular clusters, the Galactic halo contains considerable amounts of dust and ionized gas. Radio observations at the twenty-one-centimeter wavelength have detected hydrogen clouds in the halo that are moving at very high velocities.

It also includes stars that are not associated with globular clusters. For example, astronomers have found a few RR Lyrae variables both above and below the Galactic plane. The halo may also contain large numbers of faint, low-mass red dwarf stars.

The Halo's Shape

The more conventional models of our Galaxy's structure have usually described the halo as nearly spherical. The Galaxy's rapidly rotating disk of stars and dust is embedded in this spherical halo, which itself is slowly rotating. Evidence for this model came mainly from the analysis of different star counts in various directions. However, several researchers have recently found evidence that the halo is not entirely spherical. It may be flattened into an ovoid shape whose vertical axis of rotation is smaller than its horizontal axis. Different studies have come up with ratios of the halo's vertical to horizontal axis ranging from 0.8 to nearly 1.0. (An axial ratio of 1.0, of course, is a sphere).

A solution to this apparent contradiction may be found in a proposal by astronomer F. D. A. Hartwick of the University of Victoria, in Canada. Hartwick analyzed the distribution in space of metal-poor RR Lyrae stars in the halo. These are RR Lyrae stars whose [Fe/H] ratio is less than −1, or one-tenth that of the sun. He concluded that the halo has two parts. The first is a spherical outer component, and the second is a more flattened inner region, which is dominant near our sun. This inner region has an axial ratio of about 0.6—noticeably flattened.

The distribution in space of some metal-poor globular clusters fits this model of the Galactic halo. Metal-poor globular clusters that are less than twenty-five thousand light-years from the Galactic center appear to have a flatter distribution in space than those more than twenty-five thousand light-years from the center. In this model, the outer, spherical halo extends out to around a hundred thousand light-years from the Galactic center. The inner, more flattened part of the halo extends to about forty thousand light-years along the Galactic plane and to perhaps twenty-five thousand light-years along the Galaxy's axis of rotation.

If Hartwick's model is correct, it explains why different astronomers found evidence for two different halo structures. Some researchers (especially those using star counts) were seeing the more spherical outer halo. Others were seeing the inner, more flattened part. The inner halo is more flattened because of the way the Galaxy evolved. For example, the early Galaxy may have been a spherical, slowly rotating mass of gas and stars. As some of the matter began naturally collapsing into a disk, stars that were in the outer regions and had elongated orbits would tend to find their distribution in space also somewhat flattened out.

Another model for the Galaxy's formation also would result in a flattened inner halo. In this model, after the Galaxy's disk had already taken shape, the Galactic halo was formed by the gravitational capture of several small satellite galaxies. These satellites—like several present-day dwarf galaxy companions to our own—were very metal-weak. Those with orbits inclined at less than sixty degrees to the plane of the Galactic disk would have been dragged down toward the plane by dynamic friction before they were finally torn apart by the Galaxy's gravitational pull. Their debris would eventually form the flattened inner part of the halo. Satellite galaxies in orbits with higher inclinations to the Galactic plane also were eventually disrupted, but their debris produced the more spherical outer halo.

Finally, there is an alternative to Hartwick's explanation for the halo's structure. In 1983 astronomer Gerard Gilmore argued that the halo actually has two components: an outer spherical region and an inner part that is somewhat flattened and has been referred to as the **thick disk**. In terms of vertical structure, chemical properties, and the movements of its stars, the thick disk would be intermediate between the Galaxy's standard disk and the halo. However, it is probably not a structural part of the halo. Rather, it is either an extension of the Galaxy's more flattened thin disk, or of the central bulge around the Galactic core.

Drizzle and Rain

One of the more unusual discoveries about the Galaxy's halo over the last several decades has been the existence of a "rain" of gas in its northern reaches. Most recently, Laura Danly of the Space Telescope Science Institute in Baltimore has found evidence for "storms" and "drizzles" of ionized gas falling toward the north Galactic pole. It sounds like dialogue from a "Star Trek" episode ("We're a-headin' into an ion storm, Cap'n, an' I dinna know if the warp engines will hold!"), but it is real.

Since 1949, astronomers have noticed unusual behavior in the interstellar gas at high Galactic latitudes, particularly near the Galaxy's north polar region. These peculiar movements were betrayed by absorption lines in the spectra of stars located in the Galactic halo. Doppler shifts in the spectral lines revealed unusually high velocities in interstellar gas lying along the lines of sight to the stars. The velocities increased as the "z-distance" to the stars increased. "Z-distance" is the height above (or below, a negative z-distance) the Galactic plane.

In 1983 astronomer Elise Albert showed that the high velocities were seen only when looking at distant halo stars, not at nearer stars in the Galactic disk. The velocities tended to be negative—that is, they were indicative of something moving away from the observers. Most fell in the range of ten to thirty kilometers per second, although she found some velocities as high as sixty kilometers per second.

Other evidence for high velocities in halo gas came from studies of the twenty-one-centimeter radio emission from interstellar atomic hydrogen gas clouds. These clouds are usually about sixteen light-years across, contain about fifty solar masses' worth of matter, and have a temperature of about 70 K (70 degrees above absolute zero). As far back as 1966, astronomers had noticed these gas clouds in the northern Galactic hemisphere with both intermediate and high velocities. In 1973 clouds such as these with intermediate velocities were found to be located in the Galaxy's disk, not far from our solar system.

The location of the high-velocity hydrogen clouds remained a mystery, in large part because it is very difficult to directly determine the distance to hydrogen clouds simply from their twenty-one-centimeter radio emissions. The answer finally came from a product of the space age, an astronomical satellite named the **International Ultraviolet Explorer** (IUE).

The IUE was a joint project of NASA, the European Space Agency (ESA), and the United Kingdom's Scientific and Engineering Research Council (SERC). It was launched into Earth orbit on January 26, 1978, and has been providing astronomical data ever since. The satellite is essentially an ultraviolet telescope equipped with several spectrographs. One covers wavelengths of 115–190 nanometers, and another wavelengths of 180–320 nanometers. These two spectrographs have very high resolution—that is, they can "see" great detail in the ultraviolet regions of the spectrum, down to about a hundredth of a na-

nometer. Other spectrographs aboard the IUE have lower resolution (about 0.6 nanometers) but are very good for detecting
the spectra of very faint objects.

The high-resolution spectrographs aboard the IUE have
turned out to be excellent tools for studying the interstellar gas
in the Galaxy's halo. As early as 1979, astronomers using IUE
data found clouds of ionized gases extending as far out as the
Large Magellanic Cloud. The clouds had velocities of about −40
kilometers per second. Since then astronomers have discovered
gas clouds that are falling toward the Galaxy's disk at speeds of
up to 100 kilometers per second or more than 200,000 miles per
hour.

In 1985 Danly found evidence that interstellar gas was
falling from the direction of the north Galactic pole, confirming
earlier observations by other astronomers. Her most recent work
has revealed the existence of a rain of ionized gas coming from
the Galactic halo and falling into the northern Galactic pole.
Ionized gas is gas in which one or more electrons have been
stripped from the atoms, leaving them electrically charged.

Danly made her discovery by using the IUE to get some
excellent-quality ultraviolet absorption spectra from seventy
stars high in the northern latitudes of the Galaxy's halo. The
distance from Earth to the stars in her sample ranged from 137
to 56,000 light-years. Their distance from the Galactic plane
ranged from 123 to 16,300 light-years above and 326 to 42,700
light-years below the Galactic plane.

The decision to look at spectral lines in the ultraviolet part
of the spectrum was not a capricious one. Some absorption lines
in the ultraviolet are produced by electron quantum jumps in
elements that are known to be found in interstellar gas and that
can be detected in gas at extremely low density. In particular,
Danly examined the ultraviolet lines produced by neutral oxygen atoms (O_I), and singly ionized aluminum (Al_{II}), carbon
(C_{II}), iron (Fe_{II}), and silicon (Si_{II}). (The Roman numeral $_I$
following an element's abbreviation indicates that the atom has
its full complement of electrons—one for hydrogen, for example, six for carbon, fourteen for silicon, and so on. The Roman
numeral $_{II}$ means that one of the atom's outermost electrons is

gone, and the atom is singly ionized.) Danly also used high-resolution twenty-one-centimeter data from H_I clouds lying along the lines of sight to the seventy stars.

The results of Danly's work give the clearest picture yet of the movements of ionized gas in the Galaxy's halo. Significant amounts of gas with very large velocities are found in the lower halo of the Galaxy. There is a considerable difference in the behavior of ionized halo gas in the Galactic northern hemisphere and in the southern hemisphere. In the southern Galactic hemisphere, there seem to be no large systematic motions of ionized gas at all. In the north, however, large amounts of ionized gas are falling from the direction of the north Galactic pole at speeds of up to 120 kilometers per second, almost 260,000 miles per hour. The movements of the northern halo gas are very complex and are complicated by several factors. Our own solar system's movement in orbit around the Galaxy, for example, adds a small but detectable shift to the absorption lines Danly studied. Complex movements of large clouds of hydrogen gas located in the Galactic disk and lying along our line of sight to the distant halo stars also complicate things.

The best explanation for the data is that we are seeing two separate phenomena. The first is a temporary "ion storm" somewhere between 800 and 5,500 light-years above the Galactic disk. This "storm" is relatively near us. It is not really in the Galactic halo itself, but in the extended part of the Galactic disk called the thick disk. There is some evidence that such "storms" might be very common throughout the Galaxy, but no one knows for sure. The gas in the "storm" appears to be Galactic in origin, rather than from outside our Galaxy. However, there is not enough evidence to say exactly where in the Galaxy it came from.

Danly has suggested several possible origins. The gas might have been flung out of the disk by some very energetic event, such as a series of closely spaced supernovae. Or it could have been torn out of the Galactic disk by the gravity of the Large Magellanic Cloud when it passed close by during its orbit of the Galaxy. Or the gas could have been tossed up by a collision of a high-velocity gas cloud with the disk. Whatever its origin, the ionized gas is now "raining" back down into the disk.

The second phenomenon, Danly suggests, is a continual "drizzle" of cooling ionized gas located more than 5,500 light-years above the plane of the Galaxy. This "drizzle" is in all four Galactic quadrants and can be seen from all parts of the Galaxy. As much as 240,000 solar masses' worth of ionized gas could be involved in this "drizzle," with an in-fall rate of about one-hundredth of a solar mass per year. Just one solar mass is about two million trillion trillion kilograms, so this "drizzle" is a pretty massive one.

The origin of the gas in the "drizzle" is also something of a mystery. One possible source is extra-galactic. The Milky Way is on the leading edge of the Local Group of galaxies as it moves through intergalactic space toward a collection of galaxies called the Virgo Cluster. The northern face of the Milky Way is the leading face of the Galaxy. So it could be sweeping up extra-galactic gas as it and the rest of the Local Group speed toward the center of the Virgo Cluster. Alternatively, the absence of a "drizzle" in the southern Galactic hemisphere could be telling us something about the Galaxy's interaction with its surrounding environment. The Magellanic Clouds—which as we know are two small satellites to the Milky Way—are now located in the southern Galactic hemisphere. For that matter, most of the galaxies in the Local Group are also clustered in that direction. Perhaps there is some kind of interaction between our Galaxy's halo and the Magellanic Clouds or other Local Group galaxies that prevents the development of an ion "drizzle" in the halo's southern regions.

Then again, the "drizzle" and the "storm" could somehow be connected. If they are, though, the event that was theoretically their source must be extremely energetic.

The other puzzle is that the gas nearer the pole is moving more slowly than the gas that is more distant. This could be caused by magnetic effects from the Galaxy's magnetic field or by drag effects as the gas gets closer to the Galactic disk and its large load of dust and gas. The bottom line, however, is that the origin of the ion gas "drizzle" in the halo and the ion "storm" in the thick disk are still a mystery.

The halo, like the corona, is a mostly empty place. Though it contains billions of solar masses' worth of matter, it is spread out through a huge volume of space. We hop about to see some of the globular clusters up close. The tourists get some good holos to take home and show to relatives and friends.

At one point we Jump to an isolated RR Lyrae star with a period of about thirteen standard hours. It is a huge old red giant with no planets. It simply hangs there in space and pulsates. Well, actually it is not hanging in space; its orbital velocity is two hundred kilometers per second, so it is zipping along. The metaship needs to rest again, so we station-keep with the star for about a standard day. We lie about a hundred AU above its north polar region, some fourteen billion kilometers distant. Three of our skin-mounted trivids are trained on the variable. One records continually in real time, the other two in time lapses of one second per minute and one second per hour. Some of the tape will be used for promotional purposes, and some for entertainment. There is no need to do any scientific work here; a tag in the compfiles tells us that this particular star was thoroughly studied more than a century ago.

A few more tourist Jumps take us to the outer edges of the inner flattened halo. We are fifty thousand light-years from the core, lying about a thousand lights below the Galactic plane. Two more Jumps will move us into the disk.

We celebrate the just-completed leg of our journey with the Halo Dance. And then we Jump.

12
THE GALACTIC DISK

In our metaship we have now moved up to the plane of the Galaxy and lie about fifty thousand light-years from the center. Spread out before us, sweeping across most of the sky, the Galaxy looks like a disk that thickens toward its center. It is warped; the edge to our right is bent downward, while to the left of our view, the disk is bent upward. A line of huge molecular clouds arc up from the upward warp and curl back over and down toward the disk plane. The central bulge hides the Monster that we journey to see. But first we will travel through the disk that lies spread out before us.

Much of it is obscured by large patches and long streamers of dark material. These are huge clouds of dust and gas, and they are everywhere in the Galactic disk. They are star crèches. Scattered hither and yon, and visible to us only through the sophisticated sensors of the metaship, are other expanding bubbles of gas. They are the tombstones of stars. We will make some stops along the way inward and examine all these objects more closely.

We take our bearings. To our left, inward, and slightly above the plane of the disk is a dwarf G2 star, yellow-white in color. It has nine planets circling it, plus an asteroid belt and two cometary clouds. It is Sol, the ancestral home of humanity.

Our metaship takes a deep breath of vacuum-energy. We Jump.

The disk of the Milky Way is the part of a spiral galaxy with which most people are familiar. It contains most of the Milky Way's *visible* mass, about sixty billion solar masses' worth. It includes huge clouds of dust and gas, and stars with a wide range of sizes, groupings, colors, luminosities, and ages. The disk is about sixty-five thousand light-years from its outer edge to the Galactic core. Its maximum thickness at the center is about thirteen thousand light-years, with an average thickness of about one thousand light-years. The vast majority of the stars in the disk are fairly young Population I stars, from about three to six billion years old. The youngest stars are in a layer only about sixteen hundred light-years thick, which runs right along the center of the disk.

The sun and its solar system—including the Earth—are located a little bit above the central plane, about twenty-six thousand light-years from the center. The Earth's equator is tilted about sixty degrees from the Galactic plane.

This "thin disk" is in turn surrounded by a so-called "thick disk" of older, metal-poor Population II stars. The thick disk reaches to about four thousand to five thousand light-years above the Galactic plane. The thick-disk component is not all that unusual. It has been found in other galaxies as well.

The most prominent components of the Galactic disk are its spiral arms. The discovery of the Galaxy's spiral arms and the effort to map their locations constitute one of the great ongoing stories of modern Galactic astronomy. The arms were first detected by their special distribution of stars. However, the most extensive mapping of their location has been done using radio astronomy to trace out the distribution of vast clouds of hydrogen and carbon monoxide gas.

The interstellar medium (sometimes abbreviated ISM) is the matter that lies in the space between the stars of the Galaxy. It makes up about 10 percent of the Galactic mass and is largely confined to the Galactic plane and spiral arms. It includes gas, which is mostly hydrogen, clouds of dust, and clouds made of molecules.

The ISM is constantly being mixed up by the shock waves from exploding supernovae, blown about by stellar winds, and compressed by spiral density waves. The lines of force from a weak Galactic magnetic field permeate the ISM. Cosmic rays spiral along the force lines, creating a Galactic background radio emission. The disk of the Galaxy is a far from quiet place.

We now know that the structure of the Galactic disk is quite complex. And stars are far from its most important component.

Interstellar Gas

Most of the gas in the ISM is hydrogen. It tends to clump into clouds, but there is also considerable hydrogen gas between the clouds. Some of the hydrogen gas consists of neutral hydrogen atoms, some is ionized (its electron is gone, and the atom— really just a proton—has a positive electrical charge), and some is made of hydrogen molecules.

The regions of neutral hydrogen gas that lie between the stars are called **H_I regions**. Remember that H_I stands for a neutral hydrogen atom, with one electron "circling" the proton nucleus and no net electrical charge. H_I regions are interstellar clouds of diffuse, mostly neutral atomic hydrogen. A typical H_I region is about sixteen light-years across with a temperature of about 50 K. It contains around fifty solar masses' worth of gas. There may be as much as three billion suns' worth of matter in all the Galaxy's H_I regions. H_I regions emit radio waves at the wavelength of twenty-one centimeters, a frequency of 1,420.4 megahertz. By tuning in to this frequency, radio astronomers have been able to detect and map the distribution of H_I regions in large parts of the Galaxy.

Another type of interstellar gas cloud is the H_{II} region. Recall that H_{II} stands for a hydrogen atom that is ionized (its electron is missing). H_{II} regions are vast clouds of ionized hydrogen. The electrons are usually knocked away from the hydrogen atoms by ultraviolet radiation from nearby hot, young stars. Sometimes the hydrogen is ionized by supernova shock waves, x-rays, or cosmic rays. The ionized hydrogen in H_{II} regions emits radio waves at a variety of different frequencies in the infrared, ultraviolet, and visible regions of the spectrum. This makes H_{II} regions glow. They are therefore also called **emission nebulas** or **bright nebulas**.

H_{II} regions are usually six hundred light-years or less in diameter. They surround the class of hot, young stars known as O and B stars. A total of about ten million suns of the Galaxy's mass is contained in H_{II} regions.

One of the best-known H_{II} regions or emission nebulas is the Orion Nebula, which lies about fifteen hundred light-years from Earth. It is about twenty light-years in diameter. It glows from the radiation absorbed from the O and B stars within and around it. These stars have surface temperatures of 30,000 K. The ultraviolet radiation they emit ionizes the hydrogen gas surrounding them. The gas also absorbs the energy of the ultraviolet radiation and emits it at lower energies. Thus, it glows in the visible region of the spectrum.

The space between these interstellar clouds of gas is also filled with hydrogen. Some of this **intercloud medium** is of cool neutral hydrogen gas. The rest of it is much hotter ionized gas at a temperature of around 10,000 K. Ultraviolet and x-ray astronomical observations have also given us evidence for extremely hot gas in the intercloud regions. This gas has a temperature of several million kelvins, enough to quintuply ionize atoms of oxygen to O_{IV}. This is called **coronal interstellar gas**. It has nothing to do with the Galaxy's corona; the name comes from the fact that this gas is as hot as the material in the sun's outer atmosphere, or corona. At least 90 percent of the Galactic disk's interstellar space is filled with coronal interstellar gas.

Interstellar Dust

About 1 percent of the interstellar medium—the matter that occupies the space between the stars—is dust. On the average, there is about one dust grain per thousand cubic meters of interstellar space. (That's a cube with sides ten meters long, the size of a spacious living room with a very high ceiling). But that dust is very effective at blocking the light from distant objects. We can see it with our own eyes. The famous Horsehead Nebula in the constellation Orion, for example, is a dark nebula whose dust concentrations dramatically cut off the light coming from more distant stars. From our perspective, the nebula's whorls of gas and dust resemble a horse's head. The dark lanes and rifts in the Milky Way, which William Herschel thought were gaps in the star clouds, are actually regions of interstellar dust.

Other dust clouds do not look dark, but rather glow with reflected light. The best example is the nebula surrounding the Pleiades, the beautiful cluster of stars in the constellation Taurus. When looked at through even a small telescope, the Pleiades are seen to be wreathed in wisps of light. These clouds are not clouds of ionized and glowing hydrogen gas. The spectrum of the nebula does not have the H_{II} lines that would identify it as ionized hydrogen. Instead, what astronomers see is the absorption line spectrum of the Pleiades' stars themselves. It is light that is reflected by dust, the dust that makes up the clouds surrounding these stars. (Nebulas such as these are called **reflection nebulas**.)

Interstellar dust reveals its presence by the way it blocks starlight—a process astronomers call **extinction**—and by the way it scatters light waves—called **reddening**. When light passes through a cloud of interstellar dust, some will be absorbed by the particles, while other light waves will be scattered and leave the cloud in a direction different from the original. The cumulative result is extinction, the dimming of the light coming through the dust cloud.

In addition, blue light waves are more readily affected by interstellar dust extinction than red waves. The result is that more red light from a distant source reaches our eyes than does its blue light. This makes the light source—star or star cluster or nebula or distant galaxy—appear redder than it really is. This is the cause of reddening. Meanwhile, the blue light that is not absorbed eventually bounces out of the dust cloud and reaches our eyes, causing the dust cloud (a reflection nebula) to look bluish.

One of the unexpected benefits of interstellar dust to astronomy is that much of it either is transparent to infrared radiation or itself emits infrared light. This means that astronomers who use infrared light detectors can see through such clouds to objects beyond and can also see the clouds themselves by their own infrared light. Interstellar dust emits infrared light because the grains act somewhat like very tiny blackbodies. If a dust grain's temperature reaches about 100 K, it emits a lot of radiation in the infrared region.

Astronomers have been able to detect numerous interstellar dust clouds from their infrared emissions. Some well-known celestial objects turn out to look quite different. One good example is the Orion Nebula. In visible light it is a glorious sight in the belt of the constellation Orion. The group of hot young stars called the Trapezium is especially striking. In infrared light, however, the visible Orion Nebula and the Trapezium turn out to be nothing more than a tiny bubble at the front of a huge cloud of gas and dust, the Orion Molecular Cloud (OMC-1). Deep within the OMC-1 is an object that is one of the strongest infrared sources in the region. It is called the **Becklin-Neugebauer object**, in honor of the two astronomers who discovered it. It is only about 300 AU in diameter (44.9 billion kilometers) and has a temperature of 600 K. It is probably a star in the process of "turning on."

The nature of interstellar dust has long been a question among astronomers. There is still no completely satisfactory answer, but we do have some good guesses. The most common elements found in the space between the stars include hydrogen and helium (of course), as well as carbon, iron, neon, nitrogen, oxygen, silicon, and sulfur. These elements can combine into many different molecules, including water (H_2O), carbon dioxide ice (CO_2), methane (CH_4), ammonia (NH_3) and various oxides of silicon, magnesium, and sulfur. Many of these compounds form ices at temperatures below 100 K, the temperatures found inside the warmest dust clouds. They are also known to be the main constituents of the nuclei of comets. Some of the silicate molecules are known to be formed in the relatively cool atmospheres of red supergiant stars. Finally, we know from the way light is both reddened and extinguished by its passage through the Galaxy's dust clouds that the dust grains themselves must be about one micrometer in radius, about the size of a wavelength of infrared light.

In fact, we may already have samples of interstellar dust at hand right here on Earth. They are called Brownlee particles.

University of Washington astronomer Donald Brownlee has for many years been busy capturing pieces of comets. NASA owns several high-flying U2 jets (the old spy planes). Brownlee has attached to the planes special dust grain collectors that are not unlike flypaper. The U2s fly so high that they are above nearly all the Earth's atmosphere. There they encounter tiny grains of dust filtering down to Earth. These grains are thought to be sputtered off comets. However, comets are themselves thought to be made of primordial interstellar material that was left behind after the formation of the sun and its planets. So the Brownlee particles, softly raining down on Earth from the tails of comets, may actually be the first cousins of interstellar dust grains.

All of this work, both theoretical and practical, has led to a commonly accepted model of a typical interstellar dust grain. At the center of a dust grain is a core of iron, graphite, or silicate compound about 0.1 micrometers in diameter. Surrounding the core is a mantle of icy material about one micrometer in diameter. The mantle is probably a mixture of water ice, carbon dioxide ice, and ammonia and methane ices.

The dust grain cores may well be created in the upper atmospheres of supergiant stars like Betelgeuse. The spectra of these stars clearly reveal the presence of elements like titanium, carbon, magnesium, and silicon. Strong stellar winds blow the outer atmosphere into space as concentric shells of gas. As the gas streams out from the star, the temperature drops enough for the metallic and silicate elements to condense into solids and combine with oxygen to form various molecules. These particles are then blown into interstellar space by the star's stellar wind. Eventually the dust may filter into the interior of dense molecular clouds. There the particles slowly collect molecules of icy matter, which stick to their surface in chemical reactions. It apparently takes about a hundred million years for interstellar dust grains of one micrometer size to form.

Other dust grains are destroyed either by collisions with one another or by evaporation caused by absorbed radiation. Both processes go on simultaneously, and the formation and destruction of interstellar dust has reached an equilibrium point. As long as the dust grains are in the cool reaches of interstellar space, they can retain their icy mantle. When they drift into hotter regions, such as clouds of ionized hydrogen (H_{II}

regions), the mantles of ice evaporate, leaving behind the silicate or metallic cores.

The Disk's Molecular Clouds

Most of the Galaxy's dust and gas are found in the disk, and about half of this is concentrated in hundreds of **molecular clouds**. The largest are called **giant molecular clouds** (GMCs). They are usually found in close association with giant H_{II} regions, the areas of ionized hydrogen gas that surround young, massive stars. The Orion region of the sky, including the Orion Nebula, is a good example of a GMC.

A giant molecular cloud is made mostly of molecular hydrogen, along with a smattering of other molecules. It is several tens of light-years in diameter and has a mass that averages about a hundred thousand solar masses. Some GMCs contain as much as ten million solar masses of matter. The average density in a GMC is about several hundred million molecules per cubic meter. The core regions of GMCs have a temperature only ten degrees above absolute zero and densities reaching a trillion molecules per cubic meter. That sounds like a lot, but it is in fact more rarefied than the best vacuum created in Earthbound laboratories.

The discovery of the "molecular Milky Way" is in many ways the story of astronomy's increasing use of new technologies to expand our understanding of the cosmos. The first interstellar molecules to be detected were found in 1937. They include the molecule of carbon and nitrogen called cyanogen (CN), a carbon-hydrogen molecule called methylidyne (CH), and its ionized version (CH$^+$). These molecules were identified by the optical spectral lines associated with them. The real breakthrough came in the late 1960s, however, when radio astronomy made it possible to detect the presence of molecules at much longer wavelengths of light than the visible spectrum.

Radio astronomy itself began in 1932. Karl Jansky, a young employee of Bell Laboratories in New Jersey, was given the task of finding the cause of the radio interference affecting transatlantic radiotelephones. Jansky was an Oklahoma boy, born and raised in the town of Norman, about twenty miles south of

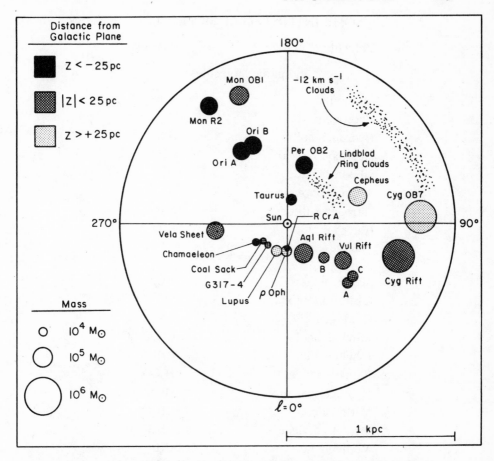

A depiction of the distribution of molecular clouds in the Galactic plane lying within 3,200 light-years (one kiloparsec) of the sun. The size of each circle indicates the mass of the cloud. The shading indicates its distance from the Galactic plane.

Illustration from T. D. Dame et al., the *Astrophysical Journal*, University of Chicago Press. Reprinted by permission of Thomas Dame and the American Astronomical Society.

Oklahoma City. He built an antenna that looked like a long linear arrangement of pipes and wires on wheels.

In 1931 Jansky discovered some radio noise at a frequency of about twenty megahertz that was clearly coming from outer space. What's more, it seemed to be coming from a particular direction in space. The signal seemed to vary not only with the time of day, but also with the month and season of the year. It

SOME INTERSTELLAR MOLECULES

Molecule	Formula
Inorganic Molecules	
Ammonia	NH_3
Hydrogen	H_2
Hydrogen sulfide	H_2S
Hydroxyl radical	OH
Silicon dioxide	SiO
Sulfur monoxide	SO
Nitrogen sulfide	NS
Water	H_2O
Organic Molecules	
Acetylene	HC_2H
Carbon monoxide	CO
Cyanogen	CN
Ethyl alcohol	CH_3CH_2OH
Ethyl cyanide	CH_3CH_2CN
Formaldehyde	H_2CO
Hydrogen cyanide	HCN
Methyl alcohol	CH_3OH
Methylacetylene	CH_3C_2H
Methylidyne	CH
Methylidyne, ionized	CH^+
Vinyl cyanide	H_2CCHCN

was acting as if it were coming from a place that was stationary with respect to the stars. By 1932 Jansky had pinned down the location of the twenty-megahertz noise. It was coming from the area in the constellation Sagittarius that astronomers suspected was the location of the center of the Galaxy.

Jansky had invented radio astronomy and he had also discovered radio waves coming from the center of the Galaxy. Few astronomers at the time were interested in this breakthrough.

One of the few people who were interested was an amateur astronomer named Grote Reber. With his own homemade radio telescope, he confirmed that the Milky Way was the source of radio emissions.

The next advance in radio astronomy was a theoretical one, and it happened in Nazi-occupied Holland in 1944. It was commonly assumed that much of the space between the stars is filled with clouds of neutral hydrogen gas. Astronomer Henk van der Hulst wrote in a scientific paper that neutral hydrogen gas might produce a unique spectral line of its own. If this should turn out to be correct, then it would be possible to detect the presence of this previously unseen component of the Galaxy. However, the rest of the astronomical community again did not take this suggestion very seriously. Like the discoveries of Jansky and Reber, van der Hulst's paper was ignored.

But that changed in 1951. That year a group of astronomers discovered what van der Hulst had first suggested six years earlier. Neutral hydrogen atoms emit radio waves at a frequency of 1,420.4 megahertz. This corresponds to a wavelength of 21.2 centimeters or about 8.4 inches, roughly the span of an adult human hand. Astronomers then realized that they could find neutral hydrogen clouds in the Galaxy.

Much of the Galaxy, especially its inner regions, is obscured by vast clouds of dust. The wavelength of the radiation emitted by neutral hydrogen is much larger than the size of the dust particles in these clouds. The radiation passes right through such clouds as if they were not there. This meant that radio telescopes could "see" through such dust clouds. Astronomers were therefore able to begin mapping the location of hydrogen clouds throughout the Galaxy. By 1954 the first large-scale maps of the Galaxy's neutral hydrogen distribution had been produced. The leaders in this project were Dutch astronomers, following in the theoretical footsteps of van der Hulst.

Until the end of the 1960s, most astronomers commonly assumed that molecules were fairly rare in the Galaxy. Hydrogen and helium are by far the most common elements in the universe. Helium almost never forms molecules, but hydrogen does. It is quite common on Earth for hydrogen atoms to form molecular hydrogen (H_2). The electrical bond between the two atoms is rather weak, however, so molecular hydrogen simply does not exist in the hot interiors of stars. Astronomers also thought it could not exist in interstellar space. Ultraviolet radiation from stars would most likely break apart any molecules of hydrogen that might form. It turns out, though, that hydrogen molecules are able to exist in interstellar space wherever they are shielded from the harsh ultraviolet radiation poured out by stars. The dust mixed in with the clouds of interstellar gas acts as just such a shield.

In fact, molecular hydrogen is *very* common in the Galaxy. It is found in the vast clouds of gas and dust that permeate the Galactic disk. These molecular clouds typically span several tens of light-years and contain several thousand solar masses' worth of matter. The largest of these clouds, the giant molecular clouds, are several hundred light-years wide and contain millions of solar masses' worth of dust and gas. The GMCs were first discovered in 1975.

Most of the gas in molecular clouds and GMCs is hydrogen. Some of it is molecular hydrogen, and some is neutral hydrogen atoms. However, these clouds also contain trace amounts of other, more complex molecules. Once again, it is radio astronomy that has detected the presence of these complex molecules by their radio emissions. For example, a radio telescope tuned to a wavelength of 35.9 centimeters will detect the presence of methyl alcohol (CH_3OH). Hydrogen sulfide (H_2S)—the molecule that causes the characteristic smell of rotten eggs—has a radio spectral "line" at a wavelength of 1.8 millimeters. And water, good old H_2O, can be detected with a radio telescope tuned to its characteristic wavelength of 1.35 centimeters. More than fifty different molecules have now been detected in interstellar space.

Interestingly enough, the discovery of these molecule-rich clouds did not result from the detection of hydrogen molecules. Molecular hydrogen is "silent" at radio wavelengths except in hot regions near young stars. Instead, the molecular clouds are most easily found by detecting carbon monoxide (CO). The second most abundant molecule in these clouds, CO was first detected in 1970 from its characteristic radio wavelength of 2.6 millimeters. CO is the standard "tracer" of molecular clouds.

Mapping the Molecular Galaxy

Radio telescopes do not focus radio waves into an image, as an optical telescope focuses visible light. Instead, they record the intensity of radiation coming from the spot in the sky where the telescope is aimed. Whatever "images" eventually result must be built up piece by piece, spot by spot. Radio astronomers call this spot the "beam." The size of the beam depends on the wavelength being observed and on the size of the antenna being used. The larger the wavelength, the larger the beam, of course. That's common sense. Conversely, the larger the dish or antenna being used, the *smaller* the beam.

Beam size determines the radio telescope's resolving power. An optical telescope with a very large mirror can detect smaller objects on, say, the moon than can an optical telescope with a small mirror. In the same way, a radio telescope with a very large antenna can "see" more detail than can one with a smaller antenna.

Carbon monoxide's characteristic radio wavelength of 2.6 millimeters is nearly a hundred times smaller than the wavelength of atomic hydrogen (twenty-one centimeters). A radio telescope tuned to CO has a small beam and therefore good resolving power. For example, at Kitt Peak in Arizona is a radio telescope with an antenna that is twelve meters (nearly forty feet) in diameter. When it is tuned to the CO wavelength of 2.6 millimeters, it has a beam width of only one arc minute. There are sixty arc minutes in one degree. Since the whole sky has a width in degrees of 360 degrees, one arc minute is a mere 1/21,600 of the celestial sphere. (Astronomers also use arc seconds as a unit of angular measurement when determining very small amounts of separation, such as the separation between the components of binary stars. One arc second is equal to 1/60 of an arc minute and, thus, 1/3,600 of a degree.)

Ironically enough, this turns out to be a problem for astronomers who want to map very large areas like molecular clouds.

Spot-by-spot mapping of a large beast like a giant molecular cloud, which may cover as much of the sky as a constellation, could take many years of work. Early radio astronomers were resolving great detail in small areas of some molecular clouds, but they were unable to get the big picture.

In 1975—the same year that the first GMCs were discovered—an astronomer at Columbia University solved this problem in a rather ingenious way. Patrick Thaddeus and his colleagues built a radio telescope dedicated to mapping the distribution of CO emissions in the Galaxy—and thus the location of molecular hydrogen clouds. This "mini" radio telescope was a dish antenna with a diameter of just 1.2 meters, or a little less than four feet—the same size as the mirror of William Herschel's great reflecting telescope of two centuries earlier. At the CO wavelength of 2.6 millimeters, its beam width was about eight arc minutes, eight times the beam width of the Kitt Peak radio telescope. The latter instrument may be useful for doing detail work on molecular clouds, but Thaddeus's mini-scope, with its larger beam width, would be better able to construct a larger picture.

The radio telescope was placed atop the fifteen-story Pupin Physics Laboratories of Columbia University, smack in the middle of New York City. It would have been ridiculous to place any optical telescope, or a radio telescope tuned to longer radio wavelengths, in the middle of Manhattan. The lights of the city and the emissions from all its radio stations would render it useless. But the Big Apple is completely "silent" at the radio wavelength of 2.6 millimeters. Thaddeus's mini could have just as well been located in the middle of the Rocky Mountains.

The 1.2-meter radio telescope at Columbia University proved to be quite good at making maps of molecular clouds like the Orion Nebula. But the process was slow, even with its eight-arc-minute beam. So Thaddeus and his team came up with an ingenious way to make its beam even larger. Instead of tracking one spot in the sky for five to ten minutes at a time, the telescope's computer was programmed to step the antenna through a sequence of sixteen different points. The points were arranged in a square made of four rows of four spots each. Each spot was separated by one-eighth of a degree, roughly one beam-width. The readings from the sixteen different spots are combined into one large "superbeam" that covers one-half a degree of the sky.

Of course, only half of the entire sky is visible from New York in the Northern Hemisphere. In 1982 Thaddeus (and Columbia University) built a second mini radio telescope, and installed it high in the Andes Mountains, in Cerro Tololo, Chile. By combining data from the twin telescopes, Thaddeus and his colleagues by 1988 put together the first complete map of carbon monoxide (and thus molecular hydrogen) distribution near the Galactic plane.

The Milky Way CO map is made from more than thirty-one thousand individual observations and covers about one-fifth of the sky. In fact, the map is actually two maps. One is an "intensity map." It shows the Galaxy as it would look to us if we could see radio waves at 2.6 millimeters. The brighter a particular region is, the more CO lies along our line of sight. The second map is a "velocity map." This is not a map in the conventional sense; it does not show the exact location of molecular clouds. Rather, it depicts the velocities of CO and molecular hydrogen clouds in various directions around the Galactic plane. The combination of velocity map and intensity map gives us a dramatic picture of our molecular Galaxy.

In terms of the distribution of stars, our sun lies about two-thirds of the way out from the Galactic center, some twenty-six thousand light-years out. However, the intensity map shows the distribution of molecular clouds to be considerably different. A few are located relatively near us, but most of the molecular material in the Galaxy is concentrated toward the Galactic center. The Galaxy is very thin in molecular clouds. The so-called inner-Galaxy ridge of molecular emissions, lying halfway between us and the center of the Galaxy (about fifteen thousand light-years away) is only about five hundred light-years thick. Its overall diameter, however, is more than fifty thousand light-years.

The velocity map also reveals some intriguing information about the disk's molecular clouds. Most of the molecular gas lying in close to the Galactic center has a very wide range of velocities. Some of the clouds are moving at velocities of around −200 to +200 kilometers per second. (A negative velocity means motion toward us; a positive velocity is motion away from us.)

This motion is not easy to interpret at first glance. For one thing, different parts of the Galaxy are rotating around the center at different velocities. The Galaxy is not like a giant rigid phonograph record, whose different parts all are moving at the same velocity. Instead, the Galactic rotation is differential, with the rate depending on distance from the center. Material closer to the Galactic center completes an orbit more quickly than matter farther away.

When we look at regions of the Galaxy closer to the center than we are, molecular clouds on one side of the center (from Galactic latitude 0° to 90°) are moving away from us and have negative velocities. They are traveling in their orbits at a velocity greater than ours. Molecular clouds in the quadrant from 270° to 0°—which are "behind" us—are moving toward us with positive velocities. Conversely, if we look at the outer Galaxy, we see that clouds in the Galactic quadrant from 90° to 180° appear to be catching up with us. That's because we are actually overtaking them, since our orbital velocity is greater than theirs. Molecular clouds in the quadrant from 180° to 270° have net positive velocities because we are moving away from them.

A simple picture of circular orbits in the Galaxy does not explain the huge velocities of the molecular clouds found exactly toward the center. All clouds in that direction should be moving at right angles to our line of sight, no matter what their distance from us. So they should all have zero radial velocity. One explanation for the great velocities detected is that a vast expansion of gas from the inner regions of the Galaxy is taking place. Another explanation, which doesn't necessarily contradict the first, is that some very massive object (or objects) at the center of the Galaxy are causing very high orbital velocities in the gas clouds of that region.

GALACTIC COORDINATES

One of the most common ways astronomers plot the location of objects in the Milky Way Galaxy is by using a Galactic coordinate system similar to latitude and longitude lines for the Earth. Galactic latitude is denoted by the letter b, longitude by the letter l. In this coordinate system, the center point is the Earth, because that's where the observing is being done from. (Yes, it is ironic but true; in the Galactic coordinate system, the Earth still lies at the center of the universe.)

The fundamental circle of the system is the **Galactic equator** or **Galactic circle**. This is the great circle on the celestial sphere that represents the path of the Galaxy. The zero point of Galactic longitude lies on the Galactic circle, along a line drawn from Earth toward the Galactic center in the constellation Sagittarius. The Galactic longitude (l) of a celestial object is its angular distance (from 0° to 360°) from the Galactic center zero point. It is measured eastward along the Galactic circle. Galactic latitude (b) is a celestial object's angular distance (from 0° to 90°) either north or south of the Galactic equator. It is measured along the great circle that passes through the object and the Galactic poles.

Galactic northern latitudes are positive numbers; southern latitudes are negative numbers. For example, the famous Orion Nebula and its associated molecular cloud, OMC-1, are located at $l = 210°$, $b = -20°$. Interestingly, the Galactic center itself turns out not to be located at $l = 0°$, $b = 0°$. The radio and infrared source Sgr A West, which is now thought to be near or at the exact Galactic center, is a few minutes of arc away. Its location is currently tagged as $l = -3.34'$, $b = -2.75'$.

The Galactic coordinate system, by the way, is not the most commonly used celestial coordinate system. That honor goes to the **equatorial coordinate system**. The fundamental circle of this system is the celestial equator. This is the great circle in which the extension of the Earth's equatorial plane cuts through the celestial sphere. The zero point is the vernal equinox—the point on the celestial equator through which the sun passes from south to north on the first day of spring. The coordinates of this system are **right ascension** (RA) and **declination** (dec), which are equivalent to longitude and latitude, respectively. Right ascension is measured in hours and minutes (twenty-four hours is a complete circle), and declination in degrees and minutes.

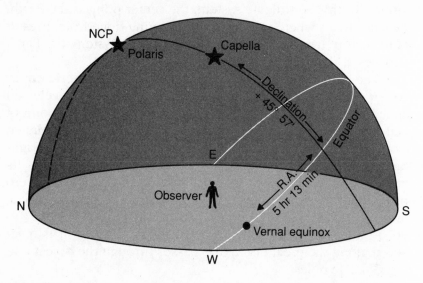

Circles in the sky showing the right ascension and declination of a star (in this case, Capella). NCP is the north celestial pole.

Illustration from Robert T. Dixon, *Dynamic Astronomy,* fifth edition, 1989. Reprinted by permission of Prentice-Hall, Inc., Englewood Cliffs, New Jersey.

The zero point of Galactic longitude in the Galactic coordinate system is located at RA 17h 42.4m, dec −28°55′ (1950) in the equatorial coordinate system. The Galactic equator or circle is inclined to the celestial equator at an angle of about 62°.

Astronomers interested in studying the Galaxy's structure have long known that the Southern Hemisphere of the Earth is the place to be. About half of the inner Galaxy (from about $l =$ 270° to 0°) is high in the sky only in the Southern Hemisphere.

Thaddeus's velocity map of Galactic CO emissions reveals that this part of the Galaxy has very few local molecular clouds compared to the Northern Hemisphere's half. In the southern Milky Way we can see bright nebulas and stars that lie many thousands of light-years distant. The half of the Milky Way visible only in the Northern Hemisphere, by contrast, is heavily obscured by the long strip of molecular clouds known as the Great Rift. These clouds lie only a few hundred light-years away from us and block most of our view of the inner Galaxy—including the nuclear bulge and the core in the constellation Sagittarius. The CO velocity map shows a nearly continuous lane of carbon monoxide emission with velocities of about ten kilometers per second—just what we'd expect from a group of molecular clouds that lie relatively near the sun.

The Warp and the Stream

As we've seen already, the Large and Small Magellanic Clouds, along with several other dwarf galaxies, are satellites of our own Milky Way. They are gravitationally bound to the Galaxy and circle it in vast elliptical orbits. The connection between the Galaxy and the Magellanic Clouds is more than just gravitational, however. It is now known that our Galaxy, like many others in the universe, is an **interacting galaxy**. It is linked to the Magellanic Clouds by a vast arc of neutral hydrogen gas called the Magellanic Stream. The interaction of the Magellanic Clouds with the rest of the Galaxy has also caused a distinct warping to the disk of the Galaxy.

The discovery of the Magellanic Stream and of the disk warp began in 1961. The famous Dutch astronomer Jan Oort decided to find out whether there was any matter falling into the Galaxy from intergalactic space. He pointed a radio telescope away from the Galactic plane and looked for signs of blue-shifted—and therefore in-falling—neutral hydrogen. Not only did Oort find what he was looking for, he found a lot of it. And he found that the in-falling neutral hydrogen was moving much more quickly than he had expected. Some of the neutral hydrogen clouds were moving toward the Galaxy at velocities of 180 kilometers per second.

Several years later, other astronomers (including Peter Wannier, who is now at NASA's Jet Propulsion Laboratory in Pasadena, California) discovered clouds of neutral hydrogen near the south Galactic pole that are moving toward the sun at one hundred kilometers per second. In 1974 came another discovery. The hydrogen clouds near the south Galactic pole are in fact part of the Magellanic Stream, the string of hydrogen gas clouds that traces a path of interaction between the Magellanic Clouds and the Galaxy. It is more than three hundred thousand light-years long and at places is a little more than thirty light-years wide. It seems likely that the Magellanic Stream was formed several hundred million years ago. As the small Galactic satellites passed close by the Galaxy, its massive gravitational force peeled vast amounts of gas out of them into a stream like a jet's vapor trail.

Meanwhile, other astronomers were trying to fit the high-velocity clouds found in the Galactic northern hemisphere into some kind of picture. In the early 1970s Gerritt Verschuur and Rodney Davies independently suggested that the edges of the Galactic disk might be somewhat warped—bent upward and downward from the Galactic plane like the brim of a jaunty hat. The northern hydrogen clouds, they said, might be parts of streamers of hydrogen gas that rise high above the Galactic plane and then curl back in. The rim of the disk on the opposite side of the Galaxy might also be warped downward.

The actual distances to the various high-velocity hydrogen clouds are poorly known. There appear to be no stars in them, including such distance-measuring stars as Cepheid variables. This makes it extremely difficult to say exactly how far away from the sun—and the Galactic disk—these clouds actually lie.

However, some work in 1988 by W. B. Burton of the Leiden Observatory has shed light on the warping of the disk. Burton made a detailed survey of the neutral hydrogen in the Galaxy's northern hemisphere. Then he combined his data with those of Frank H. Kerr, who had done a similar survey of the neutral hydrogen in the Galactic southern hemisphere. Burton then turned all the data into a movie showing the distribution of neutral hydrogen. The resulting images show distant spiral arms

THE GALACTIC DISK   237

traced out by neutral hydrogen, warping high above the Galactic plane. It is now clear that there is a distinct warp to the disk. Our Galaxy looks a little like an Australian bush hat, with the brim on one side pushed up, while the brim on the opposite side is bent slightly downward.

Recently, astronomy by satellite also has been used to detect the bends in the Galactic disk. IRAS (the Infrared Astronomical Satellite) was launched in January 1983 and operated for almost a year. It surveyed 95 percent of the sky in the infrared region of the spectrum. Among other things, it found a dust shell around the star Vega, protostars in several GMCs, and infrared emissions from comets. Astronomers have also used some of its data to find a warp in the Galaxy's stellar disk. Not only is a layer of hydrogen and carbon monoxide gas in and around the disk warped, *so is the actual disk of stars itself.* The stellar warping becomes greater at larger distances from the Galactic center, just as the H_I warp does. Some of the stars in the IRAS survey that were used to map this warping are estimated to be several billions of years old. That suggests that the warping of the disk is something that has existed for a very long time.

The warp in the disk is therefore quite real. Its cause, however, is still unknown. The Magellanic Stream was caused by a close encounter of the Magellanic Clouds with the Galaxy itself. Close flybys of these dwarf galaxies with the Galaxy would not necessarily cause such a long-lasting phenomenon as the warp. One possible cause might be an asymmetrical distribution of mass in either the Galactic halo or in the corona. That just pushes the question back one notch, since no one has any real idea of what could cause such an odd distribution of matter—either normal or "dark"—in those outer regions of the Galaxy.

The Spiral Arms

The Milky Way Galaxy is one of a class of galaxies called **spiral galaxies**, galaxies that are disk-shaped and possess two or more **spiral arms**. In this it is not unusual. More than 60 percent of the galaxies we have observed have some kind of spiral structure. Recall that astronomers divide spiral galaxies into two major classes, S and SB (the B stands for barred, for a nucleus with a roughly rectangular shape). These main categories are further subdivided into Sa/SBa, Sb/SBb, and Sc/SBc subcategories, depending on how tightly wound the spiral arms are and how large or small the nucleus is. Our Galaxy is commonly categorized as an Sb or Sc galaxy. The nucleus is either moderately sized or small, and its spiral arms are somewhat fragmented.

There is still some debate about how many spiral arms the Galaxy actually has, and even what names to give them. It's difficult to trace out the spiral structure. That's because we are actually inside the Galaxy itself. Also, dense clouds of gas and dust block our view of the Galaxy's overall structure. Other spiral galaxies can be seen to have two spiral arms winding around the nucleus. So it was long assumed that our Galaxy also has two major spiral arms. However, this may not necessarily be true. Some spiral galaxies have "arms" that are actually *segments* of arms. They appear to be arcs rather than arms. This seems also to be the case with our Galaxy.

Spiral arms in other spiral galaxies and in our own Milky Way contain many O and B stars, stars that are large, young, hot, and very blue. In addition, the overall density of matter in a spiral arm is about ten times that of the region between arms. Photos of our nearby galactic neighbor M31, the Andromeda Galaxy, taken with film sensitive to blue light, clearly show that galaxy's spiral arms. The O and B supergiant stars show up distinctly. They and the Cepheid variable stars that are young Population I stars are "spiral tracers," as it were. O and B stars are born in giant molecular clouds, so it is not too surprising that GMCs are also found concentrated along spiral arms.

Detailed mapping efforts by many astronomers over the last forty years have sketched out several sections of spiral arms. The earliest "maps" were of O and B stellar concentrations. Later radio astronomers traced out the Galaxy's spiral structure by mapping the concentration of neutral hydrogen, ionized hydrogen, and molecular hydrogen and carbon monoxide radio emissions.

Several spiral arms, or arm segments, have been traced out to more than 120,000 light-years around the disk. The sun is located on the inner edge of an arm called by some astronomers the **Orion arm** and by others the **Cygnus arm**. The **Perseus arm** lies about seven thousand light-years farther out along the Galactic plane from the Orion arm. Some six thousand light-years inward from the sun is a spiral arm called the **Sagittarius arm**. The **Carina arm** is almost certainly a continuation of the Sagittarius arm. The two are sometimes referred to together as the **Carina-Sagittarius arm**. Some astronomers think there is still another spiral arm lying about thirteen thousand light-years inward from the sun. It is called the **Centaurus arm** or the **Norma-Centaurus arm**. These arms are named for the constellations in whose direction they lie as seen from Earth. Lying still closer to the Galactic bulge are at least two other spiral arms, known as the **Scutum arm** and the **Four-Kiloparsec arm**.

The Density Wave

Spiral arms were long a puzzle to astronomers. Photos of spiral galaxies clearly showed that they existed. At first it was assumed that the material in spiral arms was somehow being physically held together and was moving along as a unit. The big question was, How? What could hold all that matter together as a unit spread over tens of thousands of light-years, and how could it all stay together for millions—perhaps billions—of years? Spiral arms would tend to "wind up." Their outer regions would be moving more slowly around the center of their galaxy than the inner regions. After a few galactic rotations, the arms should be gone. Yet everywhere astronomers looked, they saw spiral galaxies with spiral arms. None of them appeared to be in the process of winding up into nothing.

One answer to the question of how spiral arms can continue to exist was first proposed in the early 1970s by astronomer Frank Shu. It is the **spiral density wave theory**. With some modifications here and there, it seems to have emerged as the best explanation for spiral arms.

Density waves are not strange at all. We encounter them

frequently but simply aren't aware of their existence. Sound, for example, is a density wave. We open our mouths and speak, forcing air molecules in front of our face to bunch together. The first group of molecules smashes into others next to them. They in turn get pushed forward and compress the molecules in front of *them*, transferring the energy of compression to the next molecules. And so on. The density wave eventually reaches the ears of our friend, who hears us say, "Hello!"

Another good way to picture a density wave is to imagine ourselves in a helicopter some clear evening, high above the freeway called the Beltway that encircles Washington, D.C. The cars and trucks below are moving along at a steady pace near the speed limit. The headlights of the cars mark their movements and paths as they steadily circle the nation's capital. Then, suddenly, a truck in the middle lane loses one of its gears and begins quickly slowing down. Soon it is puttering along at forty kilometers per hour, the driver frantically looking for an opening in the traffic so he can pull off to the side of the freeway. Meanwhile, the cars behind it hit their brakes and also slow down. The cars begin bunching up as they await a chance to pass the ailing truck. Once they get past, they again accelerate back up to the speed limit.

What we see from the air is a curious sight. Marked by the headlights, we notice a concentrated area of cars just behind the slow-moving truck. Although the individual cars themselves move out of the denser region, the traffic jam continues to exist. It is continually replenished with new cars that come up behind the truck. What's more, the traffic jam itself is moving along the freeway. It is not composed of the same objects sticking together in some fashion. Yet it is real, and it moves—at least until the frustrated truck driver gets out of the middle lanes of the freeway.

This is pretty much how a spiral density wave works in the Galaxy to create spiral arms. Stars, gas, and dust are the "cars," and the traffic jam takes place as the density wave moves past them. The traffic jam is a region in the Galactic disk where the density of "cars" increases. We see a spiral arm.

In more detail, it works like this: As a spiral density wave moves around the Galactic disk, gas in the disk piles up at its rear edge. A shock wave forms along the rear of the wave as pressure and density in the gas increase. The compression from the shock wave causes neutral hydrogen clouds that lie in the way of the density wave to begin collapsing. They begin forming molecular clouds and GMCs. The shock from the density wave causes some of the gas to form into tiny grains of dust, creating a lane of dust along the edge of the wave. Young stars and H_{II} complexes also begin forming in the collapsing GMCs and molecular clouds. The O and B stars formed have short lifetimes, but they are also large, hot, and very bright. They clearly trace out the spiral arm. Stars of every other kind also have been formed, of course. When the O and B stars die, the others (like the sun and most of its neighbors) are left behind. Meanwhile, the density wave moves on, triggering the formation of new stars. The spiral arms continue to exist, not because they are a physically coherent structure, but because the density wave continually destroys and re-creates them.

One question not answered by this scenario is, Where does the density wave come from? It's a good question, and there is as yet no good answer to it. However, the answer may be somehow connected to the overall rotation of the Galaxy itself and its disklike shape. The differential rotation of different parts of the disk may somehow give rise to density waves.

Then there is the matter of what happens to all those O and B supergiant stars. Some of them will end their lives in spectacular fashion, violently exploding as supernovae. A large cluster of O and B stars, formed at roughly the same time by a passing shock wave, could easily *explode* at nearly the same time. The result would be a rather extended shock wave moving outward into space. The part of the group supernova shock wave moving in the same direction as the overall Galactic rotation could become the beginning of a spiral density wave.

Portrait of a Spiral Arm

It is difficult to trace out spiral arm structures using O and B supermassive stars much beyond six thousand to seven thousand light-years from the Earth. At that point the dust and gas clouds in the disk plane block out nearly all sight of these stars. One exception to this is in the Carina region of the Milky Way. This is the part of the Galaxy that lies in the direction of the southern hemisphere constellation Carina ("The Keel"). The brightest star in Carina is Canopus, an F0 II supergiant star with a visual magnitude of −0.72, lying about 195 light-years from the sun. The Carina region is the location of the Carina spiral arm. Astronomers have been able to trace it out for more than eighty thousand light-years using both optical and radio astronomy.

The existence of the Carina arm was first proposed back in 1937 by a young astronomer named Bart Bok, who later became not only a premier astronomer, but a well-respected writer of popular astronomy books and articles. Bok had observed the distribution of distant young stars in the Carina region and deduced that they seem to follow a spiral pattern. He suggested the existence of a spiral arm that runs through Carina into the Cygnus region. The audaciousness of Bok's suggestion becomes clear when we realize that only fourteen years earlier "spiral nebulas" had been shown to actually be distant star systems, not objects located inside our own Galaxy. The first twenty-one-centimeter surveys of Galactic arms would not take place for nearly another two decades.

Bok's identification of the Carina arm was later confirmed by practically every study of young Population I stars in the Carina region. For example, a 1965 survey of several hundred O and B stars showed that the stars were spread out over distances of six thousand to more than thirty thousand light-years, between Galactic longitudes of 282° and 292°. Their distribution has a sharp outer boundary, just what one might expect from a spiral arm. Maps and atlases of radio emissions and hydrogen-alpha emissions made in the 1960s also confirmed the existence of the Carina spiral arm.

By 1970 Bok and two colleagues had put together what is now the generally accepted picture of the Carina arm. This spiral arm begins at a point about 6,200 light-years inward from the sun near Galactic longitude (l) 295°. At about 283° the arm crosses the **solar circle**, the imaginary circle around the Galactic center that passes through the sun's location. At that point we see the Carina arm tangentially. The arm then continues to extend outward for about thirty thousand light-years or so, gently curving toward the higher Galactic longitudes. Astronomers have detected twenty-one-centimeter radio emissions from a lane of neutral hydrogen gas that marks the location of the most distant parts of the Carina arm.

Since the 1970 portrait by Bok and his colleagues, astronomers have continued to learn more about the Carina arm. Some of the most exciting work has been done by Patrick Thaddeus and his associates at Columbia University, using their mini–radio telescope to map the location and distribution of molecular clouds in the Galaxy. At first it was not possible for them to work on the Carina arm; it is visible only in the Southern Hemisphere, and the first mini was in midtown Manhattan. So initially Thaddeus and his team created CO maps of other spiral arms. For example, they found that the largest molecular clouds in the Northern Hemisphere clearly outline the Perseus arm, at thirty-nine thousand light-years from the Galactic center, and the Sagittarius arm, twenty thousand to thirty thousand light-years from the center.

In 1983 Thaddeus installed the sister telescope of the first mini at Cerro Tololo in Chile. It was then possible to begin mapping the Carina arm in CO emissions, and to find the location and distribution of molecular clouds in that arm. The result has been an extraordinarily detailed picture of one of our Galaxy's spiral arms. For example, astronomers have long known that an H_I layer of hydrogen in the disk gets thicker the farther one moves into the outer regions of the Galaxy. Thaddeus has found that the CO layer along the area of the Carina arm increases in thickness in the same manner. The so-called "half-thickness" of this layer of carbon monoxide gas increases from about 365 to 600 light-years.

While mapping the Carina arm using CO radio emissions, astronomers have also confirmed the warp of the Galactic disk. This takes place in the area of the Carina arm beyond the sun, between thirty-five thousand and forty-one thousand light-years from the Galactic center. The average midplane of carbon monoxide gas along that part of the arm dips *downward* from about 160 to 540 light-years below the Galactic zero latitude. The H_I layer bends downward in the same fashion and lies within a hundred light-years of the carbon monoxide molecular midplane.

There is still more, though. Thaddeus and his associates were able to catalog the location of thirty-seven molecular clouds along the Carina arm from 282° to 336°. Each cloud has a mass that averages around a million suns. The total mass of the molecular clouds *that they mapped* in the Carina arm was about forty million solar masses. The molecular clouds lie about every two thousand light-years along a spiral arm segment that is nearly eighty-two thousand light-years long. The arm has a spiral "pitch" or angle of about ten degrees.

"Pitch" is not difficult to understand. Suppose we look down on the Galaxy from above its north pole. We draw a "horizontal" line through the Galactic center. If we now draw a second line with a pitch of ninety degrees, it would be perpendicular to our first line. A line with a pitch of forty-five degrees would lie at that angle to the first line. So a spiral arm with a pitch of ten degrees moves out, away, and around the Galactic center at a continuing angle of ten degrees.

As long as we're looking down on the Galaxy from above, we can also picture the distribution of molecular clouds along the Carina arm. Thaddeus discovered that, from Galactic longitude 270° to 300°, all but two of the molecular clouds they mapped fit into the Carina arm. The same was mostly true of molecular clouds from 301° to 348°. One cloud, with a mass of forty thousand suns, was found in the Carina arm at a distance of nearly seventy-two thousand light-years from the sun.

The most fascinating discovery, however, was in the other direction. For in the other direction, lying inside the solar circle, is the Sagittarius spiral arm. Astronomers have long argued about whether the Sagittarius and Carina arms are actually one and the same. Some have felt they are; others that they are parts of other arms. One proposal, for example, held that the Sagittarius arm was the outer end of another spiral arm that lay further inward in the fourth Galactic quadrant. The Carina arm, in this suggestion, was joined to the Perseus arm. Another suggestion held that Bok's initial proposal was correct, that the Carina arm was linked to the Cygnus arm.

The data from the CO and neutral hydrogen molecular clouds have pretty much settled the issue. There is no evidence that the Perseus and Carina arms are joined together. They both have molecular clouds within them, but their distribution of mass over distance is quite different. In other words, the two arms have different molecular cloud "fingerprints."

On the other hand, the same kind of evidence reveals a striking similarity between the Carina and Sagittarius arms. What's more, the clouds in both arms fall along a single logarithmic spiral of ten degrees. The most reasonable interpretation of these data is that the Carina and Sagittarius arms are actually the same spiral arm. It has a spiraling pitch of ten degrees, is at least 127,000 light-years long, and extends *two-thirds of the way around the Galaxy*!

Further Inward

About 1.2 billion solar masses of molecular hydrogen lie in a broad region extending from about 33,000 to 6,500 light-years from the center. Patrick Thaddeus's CO survey revealed that about 70 percent of the carbon monoxide gas detected lies in a well-defined ring within this area. From its inner to outer edge, the ring itself is about thirteen thousand light-years thick. The inner radius of the ring lies about twelve thousand light-years from the center, and the outer edge is about twenty-five thousand light-years distant. The mean vertical thickness of the ring is about three hundred light-years, less than one-sixtieth its lateral width. Thus, it truly does resemble a ring rather than, say, a doughnut.

There is also evidence that this molecular ring is not perfectly flat, but rather is significantly warped in places. That would fit with other evidence that the Galactic disk as a whole is not flat, but is bent in places like the brim of a cowboy hat.

Observations in the infrared and gamma-ray regions of the spectrum have also detected a ring of matter at about this same location. The structure has been dubbed the **Five-Kiloparsec**

Ring because of its distance from the Galactic center. The ring is asymmetric. In the sky of Earth's Northern Hemisphere, it lies between sixteen thousand and twenty thousand light-years from the center. In the Southern Hemisphere it has been seen to lie between thirteen thousand and sixteen thousand light-years from the Galactic core.

Still more structure is found in the area of the Galactic disk lying inward from our location. The spiral arms we've just explored are not the only such features in the Galaxy. Two of the more prominent and better-studied spiral arms in this region are the Four-Kiloparsec arm and the Scutum arm. The Scutum arm lies between the Four-Kiloparsec arm and the Carina-Sagittarius arm. As its name suggests, the Four-Kiloparsec arm lies about four thousand parsecs or about thirteen thousand light-years from the Galactic center. It is associated with the Five-Kiloparsec Ring. The arm and the Ring in turn appear to contain the youngest regions of star formation in the Galaxy.

These two spiral armlike features are still not known well. It is difficult to map the location and features of spiral arms that lie that far inward of our Earth's position, some twenty-six thousand light-years out in the Galactic disk. Interstellar extinction from gas and dust clouds obscure the view, and the overall background radiation of gamma rays, radio waves, and x-rays makes it somewhat difficult to "see" things like spiral arms in those regions of the electromagnetic spectrum. It is fairly well accepted by astronomers, though, that these two arms or armlike segments do exist.

Still further inward, the structural character of the disk changes so much that it is no longer appropriate to speak of it as the Galactic disk. At about eight thousand light-years from the center, past the Scutum arm and the Five-Kiloparsec Ring, we enter the Galactic bulge. Our journey is coming to an end.

Inward and inward we move, Jump following Jump. We stop a while in Gould's Belt to visit the solar system. Gould's Belt is the spur of bright stars that angles off from the Orion Arc. The arc *and the belt are the location of the Old Home Star. Then it's a 1,500-light-year leap to the Orion Molecular Cloud and its nebula. The tourists take holos of the Trapezium, the B-N Object, some Bok globules, vid panoramas of the glowing gas. We hop in and out of H_I and H_{II} regions. We have lots of time, so the metaship suggests a tour around the Carina-Sagittarius arm. It's a 120,000-light-year trip, but she's pumped and full of energy. Big Jumps: We make the trip in less than two weeks.*

Then it's time to begin moving seriously inward. Our goal is the center of the Galaxy. There are other things to see, places to go. We Jump into the Five-Kiloparsec Ring. The ship catches her breath. Then inward again.

We enter the bulge. The core awaits us within.

13
THE CENTER OF THE GALAXY

We have Jumped in past the Scutum and Four-Kiloparsec arms. We stop to gaze at some of the gas clouds in the Three-Kiloparsec Arm, then Jump in still further. The metaship then slides into the Galactic bulge itself, moving along the plane of the disk.

We have seen it from afar, and we know what it looks like and how big it is. The bulge resembles a squashed sphere of tightly packed, unblinking fireflies. But the bright points of light are really stars. The closer they are to the center of the Galaxy, the more tightly packed they are. This flattened sphere of light has a total mass of some 140 billion suns. It has a radius of about eight thousand light-years along the plane of the disk. Its radius along the north-south rotational axis is about five thousand light-years. For the most part the stars are old, red Population II giants and supergiants. But there is evidence of recent star formation in some areas. There is much more than stars in the bulge. Here too are clouds of gas, much of it ionized and some of it neutral hydrogen.

We Jump again. We are fifteen hundred light-years from the center. The tour director points out a molecular cloud a few light-weeks distant. It is named Sagittarius B2, she says. It is one of many and contains only a fraction of the total amount of gas in this part of the bulge. Just the neutral gas in this region amounts to several hundreds of millions of solar masses.

At the center of the Galactic bulge is the Galactic core. It is about one thousand light-years in diameter. The next Jump moves us into the core itself. We are now a hundred light-years inward from the Expanding Ring and three hundred light-years out from the center.

248

The closer we get, the faster everything is moving. Fast, fast. It is as if the gas and dust and clouds and stars were whirling about in some gargantuan whirlpool. The radial velocities are astonishing, even to the knowledgeable in the crew and the passengers who have taken this tour before. Hundreds of kilometers per second! And more!

Our next stop will be within the innermost parsec of the Galaxy—some three light-years from the center of the Galaxy. That is as far as we dare venture—with tourists. Even in this day and age, with multiple safeguards and intelligent metaships, Anderson-Bear Cruises takes no chances.

And then—

In nearly all respects but one, the nucleus of the Galaxy appears to be fairly normal, at least compared to other spiral galaxies. The reason we can see so much detail in the inner light-year is that we are so close to it—only twenty-six thousand light-years distant. The nucleus of the nearest spiral galaxy, M31, is more than two million light-years away.

The Galactic center's brightness turns out to be quite normal in every region of the electromagnetic spectrum but one. The Galaxy's core is very bright in the far-infrared part of the spectrum. One way to illustrate this is to compare infrared images of our own Galactic core with those of two edge-on spiral galaxies, NGC 4565 and NGC 5907. The former is classed as an Sb spiral, and the latter as an Sc. The comparison indicates that our Galaxy falls somewhere between those two, but a little closer to the Sc category.

Because of this, a few astronomers doubt that there is anything unusual or bizarre at the Galactic center. The most prominent of these is George Rieke of the University of Arizona. Rieke does not believe that the Galactic center harbors a supermassive black hole. He does believe some interesting things are happening there—but in his opinion a black hole is not one of them. In this respect, Rieke and his supporters are the "loyal opposition."

It's important to have a "loyal opposition" in any scientific field. True scientific advances never take place when everyone agrees on something. They happen when someone takes exception to the prevailing view, offers alternative hypotheses and theories, and points out the nagging little exceptions to the rules. Much of the time, the minority view is incorrect. Occasionally, the dissenter turns out to be a visionary, and a new insight is gained.

However, the most recent evidence strongly suggests that the no-black-hole faction among Galactic astronomers is wrong.

The Barred Galaxy

Some spiral galaxies called **barred spirals** have arms that emerge from the ends of a bright central barlike feature that extends across the central bulge. There is some evidence that our Galaxy has a barred asymmetry to the inner part of its disk. It is in the form of what is known as the **Three-Kiloparsec Arm** and was detected as early as 1957. This is a distinct continuous Galactic feature that can be seen in twenty-one-centimeter radio emissions. It seems to lie at a distance of about three kiloparsecs or about 9,800 light-years from the Galactic center. From our perspective it lies in front of a molecular cloud in the central region called Sagittarius A (abbreviated Sgr A) and seems to be moving away from the center at a velocity of fifty-three kilometers per second.

The flow of gas in the Three-Kiloparsec Arm is not circular, and that has been something of a puzzle to astronomers. One early speculation was that the Three-Kiloparsec Arm is an expanding ring of gas created by a superexplosion at the Galactic center. Most astronomers and astrophysicists now consider this a very unlikely explanation. Such an explosion would be truly gargantuan. The total kinetic energy released by just one supernova, for example, is about 10^{51} ergs. (An **erg** is a commonly used unit of energy in scientific circles.) An explosion strong enough to create the Three-Kiloparsec Arm would require more than 10^{58} ergs, or total kinetic energy of *ten million* such supernovae going off more or less at once. Not even Larry Niven's science fiction scenario of the exploding Galactic nucleus in "At the Core" was that big.

A more plausible explanation for the gas movements of the

Three-Kiloparsec Arm is that the inner regions of the Galaxy's disk have a barlike shape. This is not a new idea. Astronomer Gerard de Vaucouleurs was suggesting it as far back as 1963. If such a barlike structure exists, as that part of the disk rotates about the Galactic center it will disrupt the flow of interstellar gas lying beyond it. This would show up most clearly as a noncircular flow of gas in the region between six thousand and thirteen thousand light-years from the center. That is where the Three-Kiloparsec Arm is located, with its apparent expansion at fifty-three kilometers per second.

At the same time, the evidence is clear that the flow of gas in the central three-thousand-light-year region of the Galactic bulge is circular. This might appear to contradict the evidence for a barlike structure in the inner disk, but it does not. For example, the barred galaxy NGC 1365 has a region of gas flow similar to that in the Three-Kiloparsec Arm region of our Galaxy but also has circular flows of gas in its bright bulge area. Because the inner regions of the Galaxy around the bulge are more spherical, the gas that lies in them will tend to follow circular streamlines in its journey around the center. The barred structure of the inner disk regions in turn will have their own separate effects on the gas in that region.

The Galactic Center at Different Wavelengths

It is not possible to see the center of our Galaxy at visible wavelengths of light. Clouds of gas and dust that lie in our line of sight to the Galactic center block the passage of nearly all light wavelengths from red through violet. Despite this, there are two different ways in which it is possible to get a hint of what the Galactic center might look like.

First, we can look at the central bulge of the Galaxy itself as it appears in the night sky. In the direction of the constellation Sagittarius, the starry band of the Milky Way becomes clotted and thick with stars. Because we are looking at it from the inside rather than from outside, the central bulge does not look like a bulge. Also, the dark clouds of dust and gas block much of it from our view. However, we can still see a hint of the immense concentration of stars that lie in the central area of the Galaxy.

A second way to get an idea of what our Galactic central region looks like is to take a look at other spiral galaxies. Galaxies like M31 (the Andromeda Galaxy), M33 (the Triangulum Galaxy), M104 (the Sombrero Galaxy), and NGC 4303 give us an idea of what our own Galaxy's central bulge may look like and a sense of the incredible concentration of stars, gas, and dust that accumulates at the Galactic center.

Of course, visible light is not the only kind of radiation that exists. Gamma rays, x-rays, infrared radiation, and radio waves differ from light waves only in their energy levels and wavelengths. These wavelengths of light penetrate the veils of dust and gas that hide the center of the Galaxy from our sight. Because astronomers have specialized instruments that can "see" in these other electromagnetic wavelengths, we can now look deep into the heart of the Galactic core.

Since the 1950s astronomers have been using radio waves to map different parts of the Galaxy, including the center. As we saw earlier, in 1932 Karl Jansky discovered that radio waves were coming from the direction of the constellation Sagittarius. At that time no one understood what that meant. No one had any idea of the actual origin and cause of this radiation. Nor did there seem to be any immediate prospect of finding out. Jansky's crude antenna array was only barely directional. In fact, it wasn't until the end of that decade that astronomers finally concluded that the center of the Galaxy lay in that direction. Besides, even if the radio waves *were* coming from the center of the Galaxy, there was no way to *see* the Galactic core. So the significance of Jansky's discovery lay hidden for many years.

Things began to change after World War II, and one major reason was World War II. That global conflict saw the rapid development of radar and microwave technologies by Great Britain and the United States. Along with that came an increasing interest in cosmic radio noise and the development of newer, larger, more sensitive, and more highly directional radio antennas. With these new instruments, astronomers would begin in the 1950s to map the Galaxy in radio-wavelength light.

Radio maps have now pinpointed the location of many giant molecular clouds in the Galaxy and have traced out segments of spiral arms. They have also presented us with a fascinating portrait of the central part of the Galaxy. It is wreathed with streamers of gas that loop up and out of the center, then back again. Narrow bands of gas some two hundred light-years long arc out and away from the central region, then back again. They appear to follow lines of magnetic force. Closer to the center, radio maps have detected gas clouds spiraling into the center, as well as three distinct tiny regions emitting radio waves. These regions are called Sagittarius A West (abbreviated as Sgr A West), Sagittarius A East (Sgr A East), and Sagittarius A* (Sgr A*).

Radio waves have become one of the two most important windows into the Galactic center. The other is the infrared region of the spectrum. Jansky's discovery of radio waves from the center of the Galaxy had been completely by accident. The infrared window opened as a result of a somewhat more deliberate search. The California Institute of Technology (Cal Tech) in Pasadena, California, has a long and prestigious history of astronomical research. Not only does it own and operate the Jet Propulsion Laboratory for NASA, it is also responsible for one of the world's largest ground-based telescopes, the famous 200-inch Mount Wilson reflector.

Gerry Neugebauer was (and still is) an astronomer at Cal Tech. In 1967 he and several graduate students were using some infrared radiation detectors hooked to a telescope to examine cool giant stars and the clouds of dust and gas that surround them. These types of stars may often be very dim visually, but they shine fairly brightly in infrared light. In particular, they were using detectors that are sensitive to infrared wavelengths at about 2.2 microns. These wavelengths of light are about two to four times as long as visible light.

One of the graduate students working with Neugebauer was a young man named Eric E. Becklin. It occurred to him that it might be worthwhile to use the 2.2-micron detector to look at the region of the Galactic center. Perhaps there might be some infrared emissions from that region. Becklin turned out to be right. The Galactic center area is extremely bright in the infrared part of the spectrum.

He and Neugebauer went on to thoroughly map the center region at 2.2 microns. They discovered that the strength of that wavelength peaks at about the point where many astronomers thought the actual center of the Galaxy was located. A wavelength of 2.2 microns is near the upper end of the infrared part

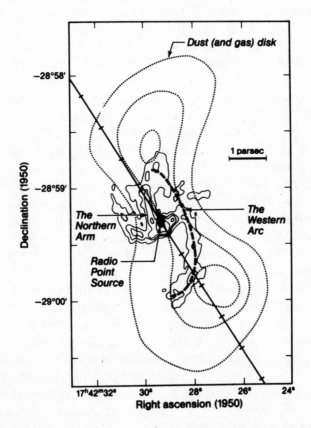

A radio map of the central twenty-six light-years of the Galaxy. The Galactic plane is denoted by the diagonal line running from the upper left to the lower right. The radio point source marked by the black dot is Sgr A*. The map also shows the "minispiral," imaged at two-centimeter radio waves, and the location of the extended dust and gas disk at the center.

Illustration from *The Galaxy*, Kluwer Academic Publishers, Inc., 1987. Courtesy of Reinhard Genzel. Reprinted by permission of Kluwer Academic Publishers, Inc.

of the spectrum, which extends down as far as about thirty millimeters.

By observing the Galactic center from the ground at certain other infrared wavelengths, astronomers have been able to actually see and photograph individual stars and clusters of stars very near the core. They have also been able to map the distribution of infrared-emitting dust clouds and streamers near the center.

Not all these wavelengths can be detected from the ground. The Earth's atmosphere, in fact, absorbs a wide range of electromagnetic waves, from the microwave up through gamma rays. This is another reason why the Galactic center has not been visible to human eyes—until recently. Orbiting observatories like the Infrared Astronomical Satellite (IRAS) and specially equipped planes like NASA's Kuiper Airborne Observatory (KAO) have enabled researchers to observe the universe—including the Milky Way's hidden center—at many different infrared wavelengths.

Because the infrared and radio windows are so important, we'll see through them in greater detail as we begin our final journey into the center of the Galaxy and confront the Monster that awaits us there. However, astronomers have also been examining the center at other wavelengths, especially with x-rays and gamma rays.

The Galactic Center at X-Ray Wavelengths

The Earth's atmosphere is nontransparent to all x-rays from cosmic sources; they never make it to the ground. The only way to detect them is to get above most of the atmosphere. So x-ray astronomy really only began after World War II, when powerful-sounding rockets and large helium-filled balloons became available.

The first detection of cosmic x-rays was in 1962. It was made by the Italian-American astronomer Riccardo Giacconi and his colleagues using a rocket that carried an x-ray detector. They were looking for x-rays that might be coming from the moon's surface. What they found was an object called Scorpius X-1, which is now known to be a binary star about nine hundred to eighteen hundred light-years from Earth in the direction of the constellation Scorpius ("The Scorpion"). A second cosmic x-ray source, Taurus X-1, was detected in 1963. By 1970 x-ray detectors carried aloft by balloons and rockets had found about thirty x-ray sources in the sky, including one associated with the powerful radio galaxy Virgo A. Taurus X-1 was found out to actually be the famous Crab Nebula, the supernova remnant of a star that blew up in the year 1054.

Instruments carried into the upper atmosphere by balloons and rockets as far back as the 1950s and 1960s detected some x-ray emission from the center of the Galaxy. However, detailed x-ray observations of the Galactic center really began with the launching of the *Uhuru* x-ray observatory satellite in December 1970. *Uhuru* (the word means "freedom" in Swahili) was launched from an ocean launch platform off the equatorial coast of Kenya. The satellite provided data that led to the first detailed x-ray sky map. Eventually *Uhuru* accumulated information on 339 different sources of x-rays in the sky, which were included in the *4U Catalogue*. *Uhuru* detected x-rays coming from a region near the Galactic center that is now called Sagittarius A, or Sgr A. At that time (about 1971), it was not possible to say whether the x-rays were coming from a single diffuse source or from three point sources. Today we know that both cases are correct.

Uhuru was followed by other orbiting observatories, including *Ariel-5*, *SAS-3* (SAS stands for Small Astronomical Satellite; *Uhuru*'s first name was *SAS-1*, and *SAS-2* was a gamma-ray observatory), *Einstein*, *Hachuko*, and *Exosat*. The Spartan-1 and Spacelab-2 laboratories carried by the space shuttle also included instruments to observe cosmic x-rays.

Apparently a number of x-ray sources in the Galactic bulge and near the Galactic center are variable; their intensity waxes and wanes. One, called A1742-294, was found by *Uhuru*. It lies about thirty arc minutes from the center of the Galaxy. Other x-ray sources near the Galactic center are called "transients." They appear for a period of about a few months and then disappear. Both the variable and transient sources are likely binary stars that include neutron star companions. Gas from the companion star spirals down onto the neutron star and explodes, causing x-rays to be released.

In 1981 the *Einstein* observatory produced the first high-sensitivity x-ray images of the Galactic center. The *Einstein* images showed A1742-294, which had first been detected by the *Uhuru* observatory. Another x-ray source found by the satellite was almost certainly Sgr A West. The satellite also detected a diffuse source of x-rays that was responsible for about 85 percent of the x radiation coming from the center.

In 1985 the Spartan-1 and Spacelab-2 laboratories were carried into orbit on separate space shuttle flights. Their x-ray detectors found a pattern of four bright sources and the one diffuse source of x-rays roughly centered on the region known as Sgr A*. One of the point sources was A1742-294. The second was a source named SLX1744-299. An x-ray source named 1E1743.1-2843 had been detected by the *Einstein* observatory six years earlier. The fourth was another source first detected by *Einstein* called 1E1740.7-2942.

All of these x-ray sources in or near the Galactic center are radiating so-called soft or medium x-rays. These are x-rays whose energy levels are between 1 and 10 keV (1,000 and 10,000 electron volts). Hard x-rays have energy levels from 10 to 500 keV.

Since the launch of the HEAO-1 x-ray observatory in 1977 and through the Spacelab-2 mission in 1985, a persistent and somewhat mysterious source of hard x-rays has been detected somewhere near the Galactic center. Scientists have had difficulty determining its exact location. However, it seems likely that it is the same object emitting soft x-rays and named 1E1740.7-2942. This x-ray source is even more intriguing, because it also appears to be the source of high-energy gamma rays.

The x-ray picture of the Galactic central region is thus an intriguing one. The central few light-years of our Galaxy are home to a source of low-energy x-rays surrounded by a patchy area of diffuse x-ray emissions. The source of the diffuse emissions is still unknown. Somewhere in this same region is a mysterious source of extremely high-energy x-rays. It is as powerful as any other x-ray source in the Galaxy. No one yet knows the exact location of this high-energy x-ray source in the central region, nor what it might be.

There are a couple of reasons that astronomers have so far had considerable difficulty in actually identifying what it might be. Until now, the resolution (or "sharpness," if you will) of x-ray images has been fairly poor. The technology of x-ray telescopes has yet to catch up with that of radio telescopes, for example, or even with the resolution of infrared-imaging detectors. Also, there is simply no way to accurately match x-ray sources with visible objects, since the intervening clouds of dust and gas block our view of the Galactic center at optical wavelengths.

The good news is that this situation will change in the next several years. NASA plans to launch a new x-ray observatory satellite called AXAF (Advanced X-Ray Astronomical Facility) by the late 1990s. It will considerably improve our view of the Galactic center in x-ray wavelengths.

The Galactic Center at Gamma-Ray Wavelengths

Gamma rays coming from the direction of the Galactic core were first detected in 1970. The Earth's atmosphere is not transparent to gamma radiation. To detect this wavelength of electromagnetic energy, instruments must be carried above the atmosphere by either balloons, sounding rockets, or satellites. Galactic core gamma rays were first spotted using instruments carried to the edge of the atmosphere by a huge balloon.

Seven years later researcher Marvin Leventhal and his associates discovered something quite intriguing about some of these gamma rays. They had an energy of about 511,000 electron volts (compared to visible light's energy of about 2 electron volts). This happens to be exactly the energy of gamma rays produced by the matter-antimatter annihilation of electrons and positrons. What's more, the intensity of the radiation coming from the center clearly indicated that an enormous amount of antimatter was being annihilated. Leventhal and others estimated it to be on the order of ten billion *tons* of positrons each second.

However, Leventhal's identification of the source's location is somewhat fuzzy. It is definitely in the *area* of the Galactic center, but the pointing accuracy of his instruments was only about eight to ten degrees in the sky, and that covers a lot of territory. The moon's angular size, for example, is about half a degree, so Leventhal's "location" for the gamma-ray source covered sixteen to twenty times the moon's angular diameter.

Over the years, the intensity of the 511-kilovolt gamma radiation from the Galactic core has waxed and waned. By 1985 it had dropped so low that the best gamma-ray detectors could no longer spot it. It was as if the matter-antimatter annihilation source had just turned off. A few years later, however, it turned on again, and the gamma rays can once more be detected coming from the center of the Galaxy.

What exactly is the source of this radiation? What could possibly produce such massive amounts of antimatter? At least two different kinds of objects could be responsible. First, a neutron star that is orbiting around a more normal star can rip away material from the surface and atmosphere of its companion. That material would then spiral down onto the surface of the neutron star. In the process, a huge amount of energy would be released, possibly enough to create large amounts of positrons.

The other possibility is a black hole. As interstellar gas and dust spirals down into a black hole, it is accelerated to almost the speed of light. This process also releases huge amounts of energy, usually in the form of gamma rays and x-rays. However, suppose an entire star were to be caught in the inexorable gravitational pull of a huge black hole. As it plunged down toward the event horizon, the star would be ripped asunder. The result would be a release of enormous amounts of energy, probably more than enough to create vast quantities of positrons. The positrons, in turn, would quickly run into the remaining normal matter. The result: positron-electron annihilation and the release of 511,000-electron-volt gamma rays.

Another gamma-ray mystery possibly associated with the Galactic core has to do with aluminum. Several astronomers at NASA's Jet Propulsion Laboratory, including William Mahoney and Allen Jacobsen, have detected gamma radiation with an energy of 1.8 *million* electron volts. This is a characteristic signature of a radioactive form of aluminum called aluminum 26. The amount of aluminum 26 involved would appear to be at least the mass of several suns.

Aluminum 26 is created only in supernovae or by extremely massive, hot, and active stars known as Wolf-Rayet stars. They are ten thousand to a hundred thousand times as bright as the sun and have surface temperatures of up to 100,000 K. By contrast, the sun's surface temperature is around 5,800 K. Wolf-Rayet stars are thought to be very young and probably represent a brief and unstable stage in the lifetime of stars. Even Wolf-Rayet stars produce only a small amount of aluminum 26. It is now suspected that most of the aluminum 26 radiation comes from a supernova remnant that is actually fairly close to the solar system but happens to lie in our line of sight to the Galaxy's center.

However, the 511-kilovolt gamma radiation is still something of a puzzle. While Leventhal and his associates could not find the source for several years, scientists using a gamma-ray detector on the *Solar Max* satellite claim they saw it annually for nine years. Other astronomers have recently suggested that Leventhal's source of these antimatter annihilation gamma rays is not at the exact center of the Galaxy, but is offset by perhaps a light-year or more. If so, then it is unlikely that a central black hole is the source of the radiation.

If not, then what? And why did some astronomers see gamma radiation coming from the Galactic center, while others did not? One likely explanation is that there are *two* sources of antimatter annihilation radiation at the core. One is a diffuse glow from the central region of the Galaxy. This is caused by the decay of elements created by supernova explosions near the center. The other is Leventhal's point source. It may be a supermassive black hole, a not-so-massive black hole of about ten thousand solar masses, or something else altogether.

The question is not likely to be answered until sometime late in 1991. Leventhal and his colleagues are planning a new series of balloon observations. Meanwhile, the Soviet Union has already launched a satellite carrying a gamma-ray detector made by French scientists. It should be able to pinpoint the variable gamma-ray source to a fraction of a degree in the sky. Finally, NASA launched the Gamma Ray Observatory satellite in April 1991. It, too, should be able to pinpoint Leventhal's mysterious gamma-ray source.

Another important gamma-ray observation took place in 1988. A team of researchers from the California Institute of Technology used a huge helium-filled balloon to launch a gamma-ray detector into the upper atmosphere above Australia. It detected a single strong source of gamma rays located less than a degree from the Galactic center. All indications are that it is the x-ray source first spotted by *Einstein* and identified as 1E1740.7-2942. If this turns out to be correct, and it is indeed at or near the Galactic center some twenty-six thousand light-years distant, then it is a powerful object indeed. At gamma-ray energies of up to 200,000 electron volts, it is one of the most luminous objects in the Galaxy.

The Lobe and the Arc

The Galactic bulge is home to at least nine "Galactic bulge x-ray sources." They lie in the central twenty degrees of the Galactic plane. They all are likely to be binary stars, in which one of the pair is a low-mass star that is shedding gas onto its companion, a neutron star. When the gas falls onto the surface of the neutron star, it explodes in a frenzy of nuclear reactions, releasing x-rays in the process.

The region lying within 1,600 light-years of the center contains more than a hundred million solar masses of neutral gas. Most of the gas in this region is confined to a relatively thin layer, with a scale height of only about 150 light-years. The density is high, about ten thousand atoms per cubic centimeter as compared to an average density of about a thousand atoms per cubic centimeter in the disk.

There are about two hundred molecular clouds in the central region of the Galaxy, and most of them are confined to this thin plane. A few others, however, are aligned almost perpendicular to the plane of the Galaxy. Star formation does take place in this region, but it is somewhat patchy and is concentrated in a very thin layer. The central region also contains interstellar dust as well as gas and stars.

A particularly striking feature of the inner part of the Galactic bulge is the **Galactic center lobe**. This is an elongated clump of ionized gas some six hundred light-years long and only a few tens of light-years wide. It projects up and out of the Galactic plane—thus its name.

The Galactic center lobe in turn is surrounded by some even longer features with similar structures. They all seem to be sticking up roughly perpendicular to the plane of the Galaxy. This might mean they were flung out from the Galactic central region by a powerful explosion. Alternatively, the Galactic center lobe and its neighbors could be streamers of gas falling into the Galactic center from some unknown outer source.

The distribution of gas in the central part of the Galaxy is heavily influenced by a strong magnetic field that is present there. This has revealed itself most clearly in some dramatic radio images of the Galactic center. They show huge arcs and filaments of gas, which extend for hundreds of light-years.

Still closer to the center than the Galactic center lobe are a collection of beautiful narrow arcs some two hundred light-years long and visible in the radio part of the spectrum. Called the **continuum arc**, these streamers were discovered in the mid-1980s by Mark Morris of the University of California, Los Angeles, Farhad Yusef-Zadeh at Northwestern University, and their colleague Donald Chance.

They used the Very Large Array (VLA) radio telescope in New Mexico. It is located in the New Mexico desert not far from the town of Socorro. The VLA is a set of twenty-seven radio telescope dishes arranged in a Y shape and mounted on railroad tracks. Each dish is twenty-five meters across. Each arm of the Y is twenty-seven kilometers long. Any or all of these relatively small dishes can be electronically linked together. The VLA can therefore effectively function as one single radio telescope with a maximum aperture diameter of twenty-six kilometers. This means that the VLA can fairly quickly produce radio maps that have very high resolution. In recent years the VLA has become one of the single most important instruments for studying and mapping the central regions of the Galaxy. Even the central Galactic light-year (the region with a radius of one light-year) is now being seen in incredible detail using the VLA.

The streamers of the continuum arc glow from **synchrotron radiation**. This is radiation caused when high-energy electrons spiral around magnetic fields and radiate photons at x-ray wavelengths as they move. The field lines are aligned along the direction of the streamers, which are perpendicular to the Galactic plane.

Although no one has yet measured the strength of these fields, it is likely their intensity is as high as one milligauss. (A **gauss** is a measure of the strength of a magnetic field.) This is about a thousandth of the strength of Earth's magnetic field at the surface of the planet.

These structures provide clear evidence that the central part of the Galaxy includes a magnetic field that is uniform, pervasive, and quite strong (for the Galaxy, that is). Otherwise, the order and regularity of these filamentary structures could not be maintained for very long.

The Expanding Ring

The inner thousand-light-year-diameter region of the Galactic bulge is the **Galactic core**. We saw earlier that the gas in the bulge tends to have a circular flow as it orbits the Galaxy. This begins to change in the core, which is home to some molecular clouds with very large noncircular motions to their orbits. Many of these clouds lie in a ring that is tilted at an angle to the Galactic plane. The ring has a radius of about 520 light-years, and it appears to be expanding at a velocity of about 150 kilometers per second and rotating at a velocity of 65 kilometers per second. If the total mass of this **expanding ring** (as astronomers call it) is ten million solar masses, then more than 10^{54} ergs of kinetic energy were needed to create its current expansion velocity.

(Not all the clouds in this region belong to this expanding ring. For example, the Sgr B and Sgr A clouds are separate from it. Most of the members of this second family of clouds lie in positive Galactic latitudes, north of the Galactic equatorial plane.)

There are several possible sources for the kinetic energy of the expanding ring, and one of them involves a black hole. Some of the clouds inside the expanding ring could be on very elongated orbits. Such a cloud will occasionally plunge right through the Galactic center—and right into any supermassive black hole that might lie there. In fact, it is possible that the Sgr A cloud is right now in the process of doing just that. Assuming there are about ten or so such clouds within five hundred light-years of the center, and each cloud has the mass of about a million suns and a radius of about sixty light-years (not outlandish assumptions), a collision of a cloud with the black hole will happen once every ten million years.

A black hole with a mass of several million suns could capture as much as five thousand solar masses' worth of matter in such a collision. The gas would form a huge accretion disk around the hole, and the mass in the disk would then flow into the black hole over a span of about a hundred thousand years. The result: the Galaxy becomes what astronomers call a **Seyfert galaxy**—a galaxy with a core that is flaring up with huge amounts of energy being released. The result would be the release of as much as 10^{57} ergs of energy. Just 10 percent of that would be more than enough to set the expanding ring in motion and also cause all the random motions of the other gas clouds in the central region.

Interestingly, it appears that such a black hole would *not* release enough energy to form the expanding ring if it were swallowing stars instead of gas clouds. A supermassive black hole with the mass of a million to ten million suns would be expected to periodically swallow stars and flare up. Several astronomers have estimated that such a star-eating episode is likely to happen once every ten thousand years or so. In such a case, the energy would be released fairly quickly, and as massive as it would be, it would add up to only about 10^{53} ergs—not enough to blow the expanding ring outward to its present position.

However, the existence of the expanding ring is not itself an ironclad proof of the existence of a supermassive black hole at the center of the Galaxy. There is still another mechanism that would cause the ring to come into existence—a starburst. A violent explosion in the central region could produce a gas ring that oscillated around a position for several rotations of the

Galaxy. The oscillation period of a gas ring at about five hundred light-years from the center is about three million years. Any burst of energy that is short compared to this period is effectively a "point explosion"—as far as the ring is concerned.

Just such an energy burst is taking place in the irregular galaxy M82. In the central region of this galaxy, approximately a hundred thousand supernovae have been popping off over the last million years, a rate of about one every ten years. This supernova outburst is the consequence of a **starburst**—a rapid and massive outburst of star formation taking place in a region of about three thousand light-years or so in diameter. Thousands of starburst galaxies were discovered by the IRAS infrared telescope satellite in 1983.

If such a burst of supernova explosions took place in the central 150 light-years of our Galaxy—even spread out over a period of a million years at a rate of one every decade—it would release the kinetic energy needed to create the expansion ring. And where would the gas and dust come from to fuel such a starburst? In fact, there appears to be more than enough gas in the central regions of the Galaxy to fuel such a starburst even today.

The Circumnuclear Disk

In the region within ten light-years of the center, radio observations reveal that much of the ionized gas is arranged in streamers. These streamers of ionized gas are somewhat patchy in structure. There are also some other bloblike clouds of plasma in this area. In the inner five to ten light-years of the core, for example, dense gas clouds appear to be spiraling down into the central region in three distinct arms, called the Northern Arm, the Eastern Arm, and the Western Arc. The clouds in these arms are rotating about the center at around 110 kilometers per second. These gas streamers either are being ejected out of the central region or are falling into it like a stream of dye spiraling into a whirlpool. Both are possible, but it is much more likely that the arms are being dragged into the center by a powerful gravitational point source.

One of the most striking features of the central region of the Galaxy is the **circumnuclear disk**. This is a disk or shell of warm, turbulent gas orbiting the Galactic nucleus at a distance of about five to fifteen light-years. The inner edge of the circumnuclear disk is ionized—radiation from within is stripping the electrons from the atoms that make it up. Inside the disk is a cavity that is almost completely clear of any gas or dust. It is as if a strong wind had blown the area clean.

Wisps of gas appear to be spiraling into the center from this circumnuclear disk. It in turn is being replenished by a long, thin stream of gas that stretches in from a molecular cloud about twenty light-years away from the circumnuclear disk. The streamer was discovered in 1989 by astronomer Paul Ho and his colleagues, working with the VLA radio telescope in New Mexico. A team of Japanese astronomers had previously seen the streamer but had not been able to track it along its entire path. Ho and his group were able to do so, measuring its movement and showing that it is connected to the circumnuclear disk.

The streamer may have been started on its journey inward by a nearby supernova explosion. Such an explosion could have knocked loose some of the gas in the molecular cloud. It turns out that there is evidence for a supernova remnant lying fairly close to the molecular cloud in question.

The circumnuclear disk contains quite a bit of heavy elements. Atoms of carbon and oxygen, ionized by strong ultraviolet radiation coming from the center of the Galaxy, have been detected mixed in with the other gases in the disk. Other ionized elements detected include singly ionized neon (Ne_{II}), doubly ionized argon (Ar_{II}), and triply ionized sulfur (S_{IV}). Ne_{II} is the most commonly found of these ions, and S_{IV} the rarest. This is because it takes a lot more energy to strip three electrons from sulfur atoms than one from neon. This in turn reveals that the ultraviolet flux is plentiful in the central region, but only moderately energetic. The most likely cause of the ultraviolet radiation is hot O and B stars with surface temperatures of 30,000–35,000 K. However, stars of this kind have not yet been found in the Galactic central region.

As we move past the circumnuclear disk and into the cavity

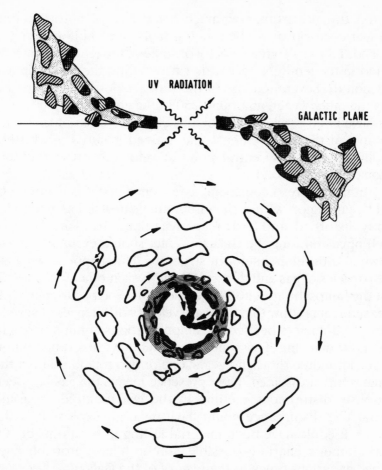

A simple but effective schematic model of the circumnuclear disk and various ionized gas streamers in the central twenty light-years of the Galaxy. In both illustrations black indicates ionized gas, heavy shading denotes warm atomic (un-ionized) gas, moderate shading marks dense molecular gas, and light shading stands for less dense gas between the clumps and streamers in the disk. The illustration at the top is an edge-on view of the disk, showing how it is warped both above and below the Galactic plane. The lower illustration is a face-on view of the circumnuclear disk. The arrows indicate the direction of motion. The circular shaded region marks the inner ring of the circumnuclear disk and the location of hot hydrogen molecular gas. This face-on view also depicts the hot shocked and turbulent inner zone and the in-falling streamers of Lo and Claussen's "mini-spiral."

Illustration from *The Galaxy*, Kluwer Academic Publishers, Inc., 1987. Courtesy of Reinhard Genzel. Reprinted by permission of Kluwer Academic Publishers, Inc.

within, things start moving much faster. The gas in the circumnuclear disk appears to be rotating at about 110 kilometers per second. This is also the velocity of some of the tenuous material found in the ten-light-year-wide cavity within the disk. Some— but not all. Several clouds in the inner region of the core are moving much faster, at around 250 kilometers per second. Others have been clocked at velocities close to 400 kilometers per second. At the very center, some ionized material is whirling about the Galaxy's central point at velocities of up to 1,000 kilometers per second.

In fact, there is considerable evidence that this central ten-light-year region within the circumnuclear disk has been violently disturbed, and fairly recently at that. Any disk or ring or shell of gas orbiting the Galactic center should eventually settle down and develop a smooth structure. This happens because the particles of gas and dust collide with each other and smooth out the lumps, bumps, and differences in velocity. Doppler shift measurements show that there are velocity differences of several tens of kilometers per second among various clumps of gas in the molecular ring. These clumps should have smoothed out in about a hundred thousand years, one or two rotations about the Galaxy at that distance. Their presence means the gas has been seriously disturbed sometime in the last hundred thousand years. The disturbance could have been an explosion at the center. It could have been material falling in from outside. Or both. Either scenario is possible, and both have probably happened at various times in the history of the Galaxy.

Finally, there is the question of the origin of the heavy elements found in both the circumnuclear disk and other areas of the Galactic center. These elements are produced only in the interiors of old stars, then ejected into space either when the stars puff away layers of their atmosphere or when they explode as supernovae. Certain rare variations (called isotopes) of some elements are found at the very center of the Galaxy. This suggests that the elements in the central region were more heavily processed inside stars than material farther out.

However, not all this matter need have come from stars that actually exist (or once existed) at the center. Some of it could well have spiraled in from regions farther out. We already know that matter is being pulled into the central part of the Galaxy. Friction and the Galactic magnetic field eventually drag in matter from the outer parts of the Galaxy.

New Stars at the Center?

Most astronomers believe there is little evidence that any star formation is currently taking place in the inner 1,600 light-years of the Galaxy. Compact H_{II} regions, as we saw earlier in the Galactic disk, are almost always associated with regions of ongoing stellar birth. The fierce radiation emitted by hot, young new O and B stars causes the surrounding cloud of hydrogen gas to be ionized, creating an H_{II} region. H_{II} clouds are extremely rare in this outer part of the Galactic core. It is possible that some star formation may be taking place in the centers of some of the larger GMCs found in that region. One such GMC is called Sgr B2, and it could be home to some new stars.

On the other hand, astronomer George Rieke argues that not only are new stars forming in the nucleus, but that it is happening right at the very center of the Galaxy. He points to the presence of a massive star, a red M supergiant dubbed IRS 7, that lies in the inner three light-years of the Galaxy. IRS 7 is losing mass at a great rate as its outer envelope of atmosphere gets blown off by a strong stellar wind. The outer edges of that envelope are being ionized by ultraviolet radiation coming from the Galactic center.

However, one red M supergiant does not a large young stellar population make. Normally the formation of such a star in a GMC is accompanied by the creation of many blue O and B supergiants. No such stars have yet been detected in the Galactic core. In addition, there would seem to be some serious difficulties with star formation in the inner five light-years of the Galaxy. This region is known to be home to *something* very massive, and it creates a strong gravitational potential in this innermost region.

Remember that gravity is like a warp in the geometry of space-time, as Einstein showed in the general theory of relativity. An extremely massive object or group of objects will cause space-time in its immediate region to "pucker" and look like a funnel-shaped well. (That's why such a region is commonly called a "gravity well.") The closer we get to the object at the bottom of this gravity well, the steeper the "walls" get. Such a strong "vertical wall" of gravity, a strong gravitational potential, would obviously make it very difficult for a molecular cloud to begin collapsing from its own internal gravity and begin creating stars.

Two other characteristics of the inner core appear to make it a very unlikely region for star formation. One is the presence of a strong uniform magnetic field. There is considerable evidence that such a field permeates the innermost regions of the Galaxy. Its field lines arc up and out of the plane of the Galaxy from both the north and south Galactic poles. Astronomers have now seen huge filaments of gas that follow the lines of force and trace them out. Finally, there is the remarkable *absence* of almost any molecular gas in the inner core.

All this would appear to rule out any chance of new stars forming at the center of the Galaxy. Yet some objects continue to perplex astronomers. What is IRS 16? A massive M supergiant? A cluster of newly forming stars? A more recent discovery has been a cluster of five infrared objects called AFGL 2004 very close to the center. In 1980 astronomers thought this cluster had two components. Later observations resolved it to three, then four separate objects. In 1990 two groups of astronomers confirmed that AFGL 2004 is made of five objects. One research team suggested that they have luminosities typical of giant or supergiant stars, but they do not appear to be either O, B, or M stars. The other team reported that the AFGL 2004 components had color temperatures of 600–1,000 K. The five objects are embedded in a thick cloud containing warm dust. They lie within four light-years of each other. The conclusion reached by both groups of researchers: AFGL 2004 is a cluster of five young stars or protostars that were recently born near the Galactic center.

If this conclusion is correct, then despite the hostile environment, it is possible for new stars to form in the center of the

Galaxy. But it may well have happened in the past. The expanding ring, as we've already seen, may have been created by a starburst event in the Galactic center a few million years ago. Even today there is enough gas and dust in the Galactic core to fuel such a burst of star making. While there is no evidence that such an event is taking place now, the presence of AFGL 2004 seems to clearly show that at least some new stars can be born at or near the center of the Galaxy.

At the Center

Over the last decade or so, the evidence for the presence of a black hole at the center of the Galaxy has steadily mounted. For example, recent infrared images confirm what most astronomers have long suspected. The distribution of stars in the central part of the Galaxy shows a strong clustering toward the center. Also, the percentage of highly luminous stars increases the closer one gets to the center. The density of stars increases toward the center and is very high indeed in the Galactic central region. The stars at the Galactic center are on the average only about five light-*days* or some 126 billion kilometers apart. That's about 1/300 the distance between the sun and its nearest neighbor, the star Proxima Centauri, which is some four light-years from us. It is clear that a very strong gravitational gradient exists in the central regions of the Galaxy. However, the strong clustering of stars toward the center is not in itself evidence for the presence of a black hole.

The most convincing evidence for many years has come from infrared and radio observations of the movements of gas and stars in the inner two or three light-years of the Galaxy. The closer to the center we look, the faster the matter is moving. Some gas clouds near the very center have been clocked at velocities exceeding a thousand kilometers per second, more than three million kilometers per hour. At that velocity, we could travel from the Earth to the moon in about seven minutes. It took the Apollo astronauts three days to make that trip.

It is not difficult to take the orbital velocity of an object and its distance from the object it is orbiting, and compute the mass of the central object. Astronomers and physicists have added up the information they have about the velocities of the stars, gas streamers, and clouds in the inner Galactic light-year. The numbers that result are pretty impressive:

There appears to be an object at the center of the Galaxy with a mass of one to four million suns.

Is it a black hole? One question to ask in response is, What would it be if it were *not* a black hole? The only possible alternative would be a large collection of extremely massive but otherwise normal stars. This is the suggestion advanced by the University of Arizona's George Rieke. We know that such stars exist, stars with masses of ten to fifty or more suns. They have very short lifetimes, measured in mere millions of years. They burn hot and fast, and they explode into supernovae at the end of their lives. They leave behind either neutron stars or stellar-mass black holes. We would need twenty thousand such stars of fifty solar masses each—ten thousand if they each had the mass of a hundred suns.

Could we fit twenty thousand such stars into a sphere with a radius of, say, one light-year? Yes. What about a sphere with half that radius? Yes again. A black hole with the mass of a million suns would be about the size of our solar system. Could twenty thousand supermassive stars fit into a sphere the size of the solar system? Possibly. But they would not stay there for very long. Within an extremely short period of time, their mutual gravitational pull would cause some to collide and others to be flung out of the region forever.

The motions of the matter in the inner part of the Galaxy strongly imply that the innermost region of the Galaxy contains a mass of several million suns. The only reasonable candidate for such an object is a supermassive black hole. Where did the matter to make the black hole come from? It could have been formed by the continual collisions of stars in the crowded central area. Or it could be the natural result of the continuing in-fall of matter from the outer Galaxy.

We know enough about the rates at which matter falls into the Galactic center from the forces of friction, gravitation, and magnetic fields to calculate how much has accumulated at the

This photo shows clouds of glowing dust and gas found in the constellation Sagittarius.
Anglo-Australian Observatory; photograph by David Malin

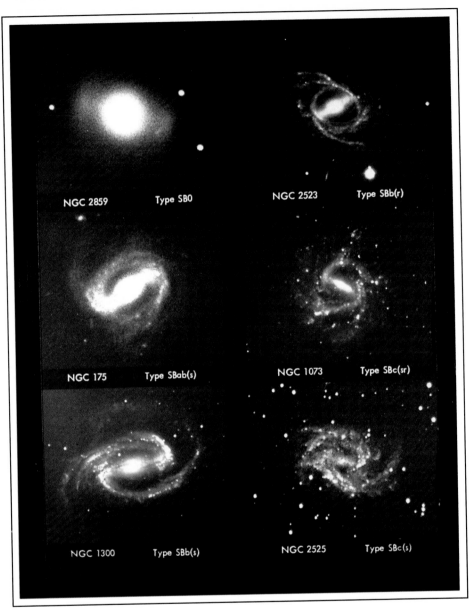

NGC 2859 Type SB0

NGC 2523 Type SBb(r)

NGC 175 Type SBab(s)

NGC 1073 Type SBc(sr)

NGC 1300 Type SBb(s)

NGC 2525 Type SBc(s)

This montage of six images shows the differences among several types of barred spiral galaxies. A barred spiral is one in which the central galactic bulge has the shape of a bar. Some astronomers believe that our Galaxy has a slightly barred central bulge.

Photograph courtesy of the California Institute of Technology

The Sombrero Galaxy is a Type Sa galaxy in the constellation Virgo, here seen edge-on. The dark band of gas and dust in its galactic disk is clearly visible silhouetted against the background of stars and its galactic bulge.
U.S. Naval Observatory

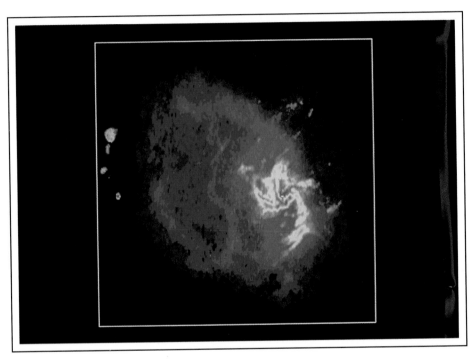

This dramatic radio image shows Sagittarius A (Sgr A), the nucleus of the Milky Way Galaxy. Much of the blue region on the left side of the picture is the region of the Galactic center called Sgr A East. It has a shell structure and may be a galactic supernova remnant about thirty-two light-years in diameter. To the right of the image is a thermal spiral known as Sgr A West.

Photograph courtesy of National Radio Astronomy Observatory/Associated Universities, Inc.; image by W. M. Goss, R. D. Ekers, J. H. van Gorkom, and U. J. Schwarz

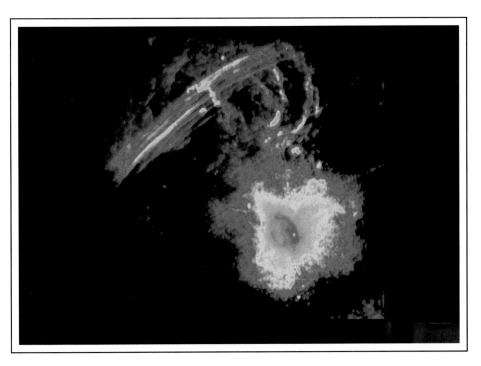

This radio image of Sagittarius A was made in 1984 using the Very Large Array (VLA). The bright blob is Sgr A, but the astonishing part of the image is the Continuum Arc, a series of filamentary structures that arch up and out of Sgr A. The filaments of the arc are each about three light-years in diameter, are perpendicular to the Galactic plane (which runs from top left to bottom right at an angle of about 60°), are parallel to each other, and are unbroken in appearance. Also apparent in this image is a halo, seventy-eight light-years in diameter, surrounding the Sgr A East shell. This halo is made of a number of large-scale protrusions that are also perpendicular to the Galactic plane.

Photograph courtesy of National Radio Astronomy Observatory/Associated Universities, Inc.; image by F. Yusef-Zadeh, M. R. Morris, and D. R. Chance

This infrared image shows the central 150 light-years of the Galaxy. In optical images this picture would be completely blank, the Galactic center obscured by clouds of dust and gas. Infrared, however, is able to pass through the clouds and be detected on Earth with special instruments. The plane of the Galaxy extends from upper left to lower right. The bright region in the center is the Galactic nucleus, now believed to harbor a supermassive black hole.

National Optical Astronomy Observatories; image by Dr. Ian Gatley and Dr. Dick Joyce

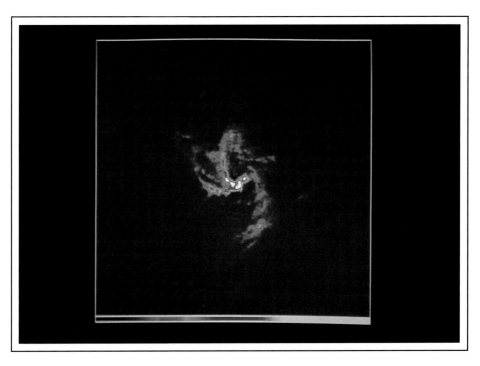

This radio image of Sgr A West has had the nonthermal radio emission subtracted, leaving a clear image of the thermal spiral at the center of the Galaxy. The three arms of this spiral are called the Northern Arm (top), the Western Arc (right), and the Eastern Arm (left). The Western Arc is part of the central molecular disk surrounding the Galactic center, seen by us at an angle. The Eastern and Northern Arms appear to be made of ionized gas that is falling into the central region. At the center of this structure, not visible in the radio image, is an infrared object known as IRS 16. It lies extremely close to the actual physical center of the Galaxy and may mark the location of a supermassive black hole.

Photograph courtesy of National Radio Astronomy Observatory/Associated Universities, Inc.; image by K. Y. Lo and M. J. Claussen

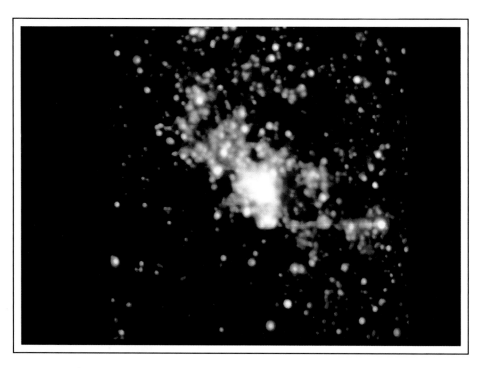

This infrared image of the Galactic center reveals the innermost region of the Galaxy. The image's diameter corresponds in size to about fifty light-years. The dark red patch to the right of the bright central nucleus is part of a circumnuclear ring of gas and dust that surrounds the nucleus at a distance of only six light-years. This image, made in 1989, is one of the most recent of the Galactic center.

National Optical Astronomy Observatories; photograph by Ian Gatley center over the lifetime of the Galaxy. It appears to accumulate at a rate of about one solar mass every thousand years. Astronomer Mark Morris thinks the inflow rate may be even higher—one solar mass every *ten* years (although much of that would also come back out in a vast stellar wind). The Galaxy is now estimated to have been in existence for several tens of billions of years. Let's assume, though, that it has been in something like its present state for about five billion years. If material has been accumulating at the center for all that time at a conservative rate of one sun per thousand years, then the total in-fall has amounted to about five million solar masses. That is very close to the present estimate for the mass of the supermassive black hole that may lie at the center of the Galaxy.

There is still another possible explanation, however, for what may lie at the Galactic center. It may not be one large black hole but rather many smaller ones. A hundred black holes, each with the mass of ten thousand suns, would also account for the present evidence of a single supermassive object at the center. Each of these black holes would have a radius of about thirty thousand kilometers. They would all be orbiting one another at the center of the Galaxy. With their relatively small size, they would be unlikely to collide. Or perhaps they have, and do. Perhaps there were once many small black holes at the center of the Galaxy, each with the mass of a few tens of suns. Over the past five to ten billion years, their occasional collisions would have resulted in a smaller collection of more massive black holes.

Finally, the "anti–black hole camp" may be correct, but not in the way they think. Astronomers and physicists today talk of exotic and mysterious things like dark matter and cosmic strings—the latter being long linearlike "flaws" in the fabric of space-time left over from the earliest instants of the universe. Perhaps the Monster at the center of the Galaxy is a dense cluster of dark matter or a tangled knot of cosmic string. It was Shakespeare who noted that there are more things in the heavens than are dreamt of in our philosophical musings. He may have been more right than *he* ever dreamed.

On the other hand, it was the philosopher William of Occam who cautioned that "entities should not be multiplied unnecessarily." In other words, the simplest of competing theories should be preferred to the more complex ones, and unknown phenomena ought first be explained in terms of known quantities. Black holes aren't exactly "known quantities," but these days they are at least more known than dark matter and cosmic strings.

The next question is, Where is it? Most astronomers have felt that there are two likely objects or locations for the black hole. One is the bright infrared object named IRS 16, which lies quite close to the geometric center of the Galaxy. Recently IRS 16 has been resolved with infrared detectors and shown to consist of at least five separate point sources of infrared radiation. It could be a cluster of new stars similar to AFGL 2004. That doesn't mean it is not the home of the black hole, since the black hole could well be lurking among them.

And the other possible home of the supermassive black hole? The three regions at the very center of the Galaxy that are power emitters of radio waves. Two of these regions are Sagittarius A West (Sgr A West) and Sagittarius A East (Sgr A East). The most intriguing radio source at the center is an extremely tiny region dubbed Sagittarius A* (Sgr A*). Sgr A* is an intense source of radio waves, its output is nearly constant, it is easily less than a light-year in diameter, and it does not seem to be moving at all in relation to anything else at the core. Many astronomers who study the Galactic center have pointed to Sgr A* as the most likely candidate for a supermassive black hole at the center of the Galaxy.

But which location is correct? The answer, along with some very intriguing evidence for the very existence of the central black hole, may have come in June 1990 from two of the astronomers who discovered the continuum arc.

Mark Morris and Farhad Yusef-Zadeh, along with their Australian colleague Ron Ekers, analyzed detailed radio images made with the VLA radio telescope. The three astronomers discovered at least seven massive blobs of ionized gas (or plasma) that appear to be spiraling out of the center of the

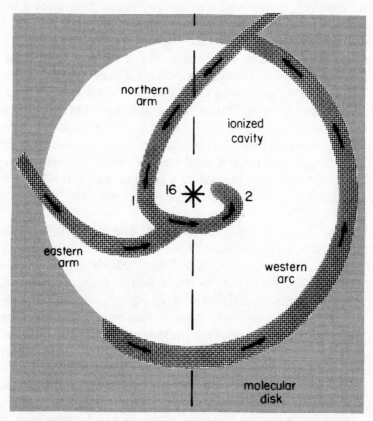

This drawing shows some of the structure of the inner few light-years of the Galactic center, particularly the structure seen in Lo and Claussen's radio image of Sgr A West. The northern and eastern arms are streamers of gas falling into the center. The western arc is not an "arm" but part of the edge of the central molecular disk of the Galaxy. The disk is tilted with respect to the Galactic plane, and the intersection is shown as a dashed line. The 16 marks the location of IRS 16, the bright infrared object at the center of the Galaxy. The 2 is a structure that is actually an extension of the northern arm, hooking around and in toward the Galactic center.

Illustration from *The Center of the Galaxy*, Kluwer Academic Publishers, Inc., 1989. Reprinted by permission of John H. Lacy, the International Astronomical Union, and Kluwer Academic Publishers, Inc.

Galaxy. The point of origin for this gas: Sgr A*. Yusef-Zadeh believes that a million-solar-mass black hole is the only explanation for this phenomenon. Of course, nothing can escape a black hole after falling past its event horizon. As we have already seen, however, matter and energy can escape from near a black hole if they have enough velocity. Yusef-Zadeh explains the seven fleeing plasma clouds this way:

The black hole at the center of the Galaxy is pulling in matter from its vicinity with its powerful gravitational field. As the matter gets closer and closer to the black hole, it begins piling up into a fast-spinning accretion disk. The black hole's gravity is pulling in matter so fast, however, that the matter becomes intensely heated as it piles up onto the accretion disk. The combination of intense heat and the disk's rapid spin causes some of the gas, now heated to a plasma, to be flung off into space at velocities great enough to escape the black hole's gravity.

Yusef-Zadeh and his colleagues have also found another intriguing piece of evidence in the new VLA radio images. The red giant star dubbed IRS 7, which lies near the Sgr A* region at the center, has a plasma tail being blown away from it. The star's plasma tail points directly away from Sgr A*, much as a comet's tail always points away from the sun. And the reason, says Yusef-Zadeh, is probably the same. The Monster at the center is emitting so much energy that the energy blows out from the center like a cosmic wind. That wind is blowing a plasma tail off the surface of IRS 7 the same way that the sun's solar wind blows a tail off the surface of a comet.

With the recent VLA observations in hand, Yusef-Zadeh is confident that he has pinpointed Sgr A* as the location of the Galaxy's black hole. (His colleague Morris has remarked that he wouldn't bet the farm on that, but he still gives it more than a 50 percent chance.) IRS 16 is not completely ruled out. However, the plasma tail on IRS 7 and the trajectories of the plasma blobs all point directly to Sgr A*. Sgr A* lies near the true dynamic and geometric center of the Galaxy. And it apparently is the location of the supermassive black hole at the core.

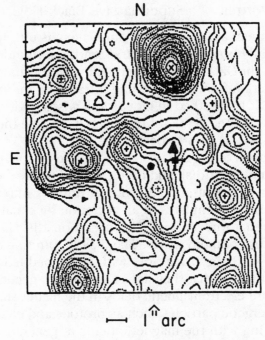

⊢——⊣ 1 arc

■ Sgr A* compact
non-thermal source

*Still another way of "looking" at the Galactic center, a region that cannot be
directly seen by the naked eye in visible light. This is a plot of equal-surface-
brightness contours based on a 2.34-micron infrared image of the central part of
the Galaxy. This map is analogous to a terrestrial topographic map that shows
different elevations on the Earth's surface. In such a map, each contour line
connects points that are the same height above sea level, and the contour lines
follow equal intervals. In this map of the Galactic center, the contour lines connect
points of equal infrared brightness. The interval between contour lines is about 0.2
magnitude. Every other contour line is dotted. The + symbols mark the location of
objects that are "bright" at the two-micron level of infrared radiation. The X near
the top (north) marks the location of IRS 7. The filled triangle is IRS 16NW. The
heavy dot is IRS 16 proper. The filled square box with the horizontal and vertical
error bars, just below the triangle of IRS 16 NW, is the estimated position of the
radio source Sgr A*. This map clearly illustrates that IRS 16 and Sgr A* are not the
same object. The supermassive black hole at the center, if it does in fact exist, may
be located in or near either object.*

Illustration from *The Galaxy*, Kluwer Academic Publishers, Inc., 1987. Courtesy of Reinhard
Genzel. Reprinted by permission of Kluwer Academic Publishers, Inc.

Portrait of a Supermassive Black Hole

The environment at the center of our Galaxy is utterly spectacular and totally bizarre. The central tenth and hundredth of a light-year of our Galaxy is crammed with electromagnetic fields, flows of radiation from the radio and infrared up through x- and gamma rays. It is sheathed in clouds and wisps of dust and gas. The result is that the center of the Galaxy must be the scene of lightning bolts and auroras beyond belief.

Several arms of gas and dust are spiraling in from the region of the circumnuclear disk and being sucked up by the Monster at the center. The matter does not fall directly into the black hole, but rather accumulates in an accretion disk that encircles the hole's equatorial region. The accretion disk is emitting powerful bursts of radio waves. The disk is spinning extremely rapidly, and so is the black hole with its embedded magnetic fields. The result is that they act as a dynamo, generating still more powerful electromagnetic fields in the matter surrounding the core. Energetic particles such as protons and electrons are also interacting with the magnetic fields to generate immense streams of radio waves.

The accretion disk emits more than just beams of radio energy. It is spinning so fast and has so much accumulated energy that several blobs of ionized gas are flung off from it. So matter falls into the disk while some of it escapes—at least temporarily.

But not all of it. The matter at the inner edges of the accretion disk continues spiraling inward, further and further inward. If the black hole is only a few hundred suns in mass, as some astronomers suggest, it has a radius of about a thousand kilometers. If it has a mass of a million suns, as other theorize, then the black hole is not even as large as our solar system. Its "radius" is somewhere between that of the sun and the orbit of Jupiter. The radius is not from a central point to a surface, of course, but rather from the point where the matter that falls into the black hole crosses its event horizon to the singularity at the center.

What is the "internal" structure of this Monster at the center of the Galaxy? It is spinning very fast, but not fast enough to become a naked singularity. So it is a Kerr black hole. First

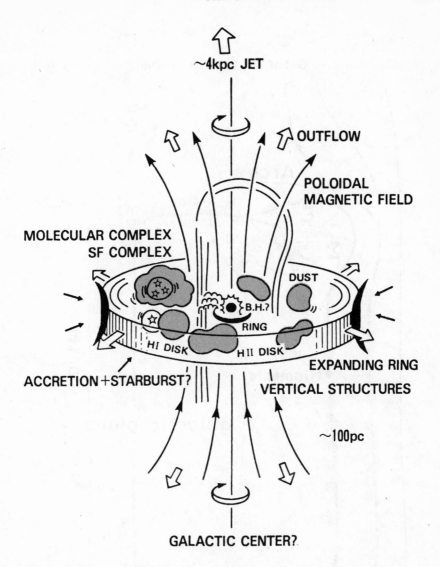

A schematic view of the Galactic center as suggested by Yoshiaki Sofue of the University of Tokyo. A variety of objects and forces are at work at the center. They include the Galaxy's magnetic field; an outflow of matter along the northern axis, including the Galactic Center Jet (or Four-Kiloparsec Jet); expanding rings of dust and gas; various molecular clouds; and, at the center, a black hole.

Illustration from *The Center of the Galaxy*, Kluwer Academic Publishers, Inc., 1989. Reprinted by permission of Yoshiaki Sofue, the International Astronomical Union, and Kluwer Academic Publishers, Inc.

GALACTIC CENTER MODEL

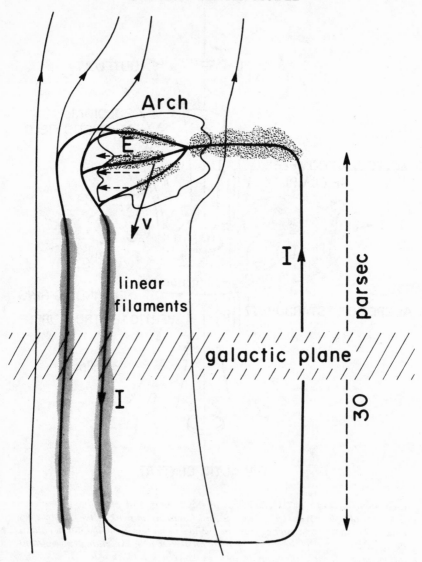

comes the outer static limit, then the outer event horizon. The region between is the outer ergosphere. Matter that falls into this region can still escape, but only with difficulty. For every second the matter remains in the ergosphere, immensities of time will have elapsed in the outside universe. Past the outer event horizon is an inner static limit and inner event horizon, with its own ergosphere. Past that point there is no return (probably).

And at the very "center" of the black hole, and thus at the center of the Galaxy, is the singularity. Since the hole is spinning, the singularity has a ring shape. Is it a gateway to other space-time continua, to other universes? No one knows. The matter that falls into the black hole cannot tell us, for no information can come out of a black hole.

This, then, is the center of the Galaxy, the center of our "island universe." This is *our* cosmogram, our vision of what lies at the center of our contemporary mythology.

We have reached our journey's end. We can go no further.

Except on the wings of imagination . . .

One of the most recent attempts to explain what is happening at the center of the Galaxy is offered by astrophysicist (and science fiction writer) Gregory Benford. Along with several other researchers, Benford suggests that electrodynamic forces are playing a major role in creating the structures near the Galactic center. This illustration depicts a "circuit" with electrical currents that may be flowing along magnetic field lines near the center of the Galaxy. The illustration covers about one hundred light-years. The center of the Galaxy is marked by the X. The lightly shaded regions are molecular clouds that are emitting radio waves. The lines with arrowheads are lines of the Galactic magnetic field. A partly ionized molecular cloud (the location of the Arch structure) moving with a velocity v drives current (I) through a circuit formed by Galactic magnetic fields. The current in turn has an inductive field (E). Essentially, this current is created in much the same way as a typical mechanical dynamo creates an electric current. The "electrical circuits" so created are more than three hundred light-years long!

Illustration from the *Astronomical Journal.* Courtesy of Gregory Benford. Reprinted by permission.

Something has gone wrong. Terribly wrong. For some reason that our shipboard metapsychiatrist cannot yet determine, the metaship had a "seizure." There is no other way to describe it.

"I don't understand," she says plaintively. "One moment I was talking with the captain, probing the accretion disk with some sensors, rebalancing the force shields, and playing a game of fairy chess with that rather obnoxious retired politician—you know, the one with the dyed hair. And the next thing I remember is seeing—seeing—seeing flashes of light in front of me. I could smell something. It was like burned insulation. And then I woke up. And here we are.

"I'm very sorry this has happened, Dave."

"That's fine," replies Dave (the metapsychiatrist). "I understand. Don't worry. We'll figure it out. Now rest your conscious modalities and allow your automatics to continue their work."

The consequences have been beyond terrifying. During the two hours the metaship was unconscious, we were captured by the gravitational pull of the black hole at the center. Fortunately, we had just Jumped to a position about three light-years above the north pole of the object. We did not fall into the accretion disk.

But we did fall into the black hole. During her seizure, the metaship had two strong "convulsions." Her convulsions were Jumps. We ended up only a few light-hours away from the hole's static limit. Fortunately, the metaship swiftly recovered her faculties after her seizure. We could not completely escape the black hole's gravitation without first passing into the hole's ergosphere. In a brilliant display of navigation, the metaship guided us into the outer edges of the ergosphere. In a matter of a few seconds, she gained enough added energy so that, along with full thrust from the Forward/Davis drive, she was able to kick herself (and us) out of the ergosphere and away from the hole. We are now moving away from the center of the Galaxy, heading south along its rotational axis.

And all the star patterns are wrong. I know. I'm the ship's astronomer. I know the patterns. They are all wrong. And I knew at once what had happened. I have run the computer simulations, moving the stars forward in time to find the matching patterns. There is no doubt. We have come forward in time ten million years.

Our world is gone. Our society. Our culture. Our mythology. What awaits us is a mystery.

We head to see what it is.

Appendix A
Scientific Notation, Including Abbreviations and Prefixes

Scientific notation is a numerical shorthand used by scientists. It is a system in which numbers are expressed as a number between 1 and 10 multiplied by a power of 10. A power of 10 is the number 10 multiplied by itself X times, expressed numerically as 10^X, where X is the number of times 10 is multiplied by itself. Thus, 10^2, or 10 squared, is 10×10, or 100, or 1 followed by two zeros. Likewise, 10^5, or 10 to the fifth power, is 100,000, 1 followed by five zeros. So, in scientific notation, the number 9,460,000,000,000 km (nine trillion four hundred sixty billion kilometers) would be written as 9.46×10^{12} km. This happens to be the number of kilometers in one light-year.

SCIENTIFIC NOTATION AND ABBREVIATIONS

Abbreviation	Prefix	Meaning	Number	Scientific Notation
T	Tera-	Million million (trillion)	1,000,000,000,000.0	10^{12}
G	Giga-	Thousand million (billion)	1,000,000,000.0	10^{9}
M	Mega-	Million	1,000,000.0	10^{6}
k	Kilo-	Thousand	1,000.0	10^{3}
m	Milli-	Thousandth	0.001	10^{-3}
μ	Micro-	Millionth	0.000 001	10^{-6}
n	Nano-	Thousand millionth	0.000 000 001	10^{-9}
p	Pico-	Million millionth	0.000 000 000 001	10^{-12}
f	Femto-	Quadrillionth	0.000 000 000 000 001	10^{-15}

Appendix B
The International System of Units

This book uses whenever possible the SI (*Système International*) units of measurement. This is the most common measurement system in the world and the one officially adopted by the scientific community. SI units are based on powers of 10 (see Appendix A), and a prefix (kilo-, milli-, etc.) for the name of the unit indicates the power of ten to use.

The tables in this appendix list the seven basic SI units of measurement, several other SI units, and SI units of length and mass that are frequently used in astronomy and physics. The third table also gives their equivalents in the English system of measurement, which is still commonly used in the United States and Great Britain.

THE SEVEN BASIC SI UNITS

Quantity	Name	Symbol
Amount of substance	Mole	mol
Electric current	Ampere	A
Length	Meter	m
Luminous intensity	Candela	cd
Mass	Kilogram	kg
Temperature	Kelvin	K
Time	Second	s

SOME OTHER IMPORTANT SI UNITS

Quantity	Name	Symbol
Electric potential difference	Volt	V
Energy	Joule	J
Force	Newton	N
Frequency	Hertz	Hz
Power	Watt	W

EQUIVALENT UNITS

International System (SI)		English System

Length

1 angstrom (Å)	= 0.000 000 000 1 m	
1 nanometer (nm)	= 0.000 000 001 m	
1 micrometer (μm)	= 0.000 001 m	
1 millimeter (mm)	= 0.001 m	= 0.03937 in
1 centimeter (cm)	= 0.01 m	= 0.3937 in
1 meter (m)	= 1.0 m	= 39.37 in
		= 3.28 ft
1 kilometer (km)	= 1,000.0 m	= 0.6214 mi

Mass

1 milligram (mg)	= 0.001 g	
1 gram (g)	= 1.0 g	= 0.002046 lb
		= 0.035 oz
1 kilogram (kg)	= 1,000.0 g	= 2.2046 lb

Appendix C
Temperature Scales

Astronomers and physicists commonly measure temperatures on the kelvin (K) scale. The temperature at which all movement of atoms and molecules would completely cease, called absolute zero, is 0 on the kelvin scale. On this scale the temperature at which water passes from a liquid to a solid equals 273.15 K. Water boils at 373.15 K.

The Celsius (C) scale is still used by many scientists, although it is no longer the official temperature scale for the scientifically accepted SI system of measurement. The degree increments on the Celsius and kelvin scales are the same; that is, $1° C = 1 K$. However, there is a difference of 273.15 degrees between the two scales' assignment of temperatures for the freezing and boiling point of water. In the Celsius scale, water freezes at 0° and boils at 100°. Celsius temperatures can therefore be converted to Kelvin temperatures by adding 273.15; Kelvin temperatures can be converted to Celsius by subtracting 273.15.

In the United States the Fahrenheit scale is the temperature scale commonly used by most people. The Fahrenheit scale was originally defined by setting the lowest temperature that could be obtained with a particular mixture of ice, water, and salt as equal to 0°, and the temperature of the human body as 96°. Later this scale was redefined so that the freezing point of water was 32° and the boiling point of water was 212°. The degree increments of the Fahrenheit scale differ from those of the Celsius and kelvin scales. To convert Fahrenheit to Celsius, one must use the following equation:

$$T_C = 5/9(T_F - 32)$$

where T_F is the temperature in Fahrenheit and T_C the temperature in Celsius. To convert Celsius temperature to Fahrenheit, use this equation:

$$T_F = (T_C \times 1.8) + 32$$

TEMPERATURES IN DIFFERENT SCALES

Description	Kelvins	Degrees Celsius	Degrees Fahrenheit
Center of the sun	1.5×10^7	1.5×10^7	2.7×10^7
Center of the Earth	1.6×10^4	1.6×10^4	2.9×10^4
Visible surface of the sun	6,000	5,727	10,340
Boiling point of gold	2,532	2,805	5,081
Melting point of gold	1,336	1,063	1,945
Fireplace fire	1,100	827	1,521
Surface of Venus	644	371	700
Boiling point of mercury	630	357	675
Boiling point of water	373.15	100.00	212
Human body	310	37	98.6
A normal room	293	20	68
Freezing point of water	273.15	0.00	32
Melting point of mercury	234	−39	−38
Freezing point of carbon dioxide	195	−78	−108
Polar cap of Mars	145	−128	−198
Average temperature of Saturn	127	−146	−231
Gasoline freezes	123	−150	−238
Nitrogen freezes	67	−206	−339
Hydrogen liquifies	21	−252	−422
Hydrogen freezes	14	−259	−434
Helium liquifies	4.2	−269	−452.11
Absolute zero	0.00	−273.15	−459.67

Appendix D
Some Important Values of Astronomy, Astrophysics, and Physics

PHYSICAL CONSTANTS AND VALUES

Electron mass (m_e)	9.10956×10^{-31} kg
Gravitational acceleration, Earth (g)	9.806 m/s^2
Gravitational constant (G)	6.672×10^{-11} N $-$ m^2 kg^{-2}
Hydrogen atom mass (m_H)	1.6735×10^{-27} kg
Neutron mass (m_n)	1.6749×10^{-27} kg
Planck's constant (h)	6.62618×10^{-34} J $-$ s
Proton mass (m_p)	1.6726×10^{-27} kg $= 1{,}836.1 \; m_e$
Speed of light in a vacuum (c)	$299{,}792$ km/s

ASTRONOMICAL VALUES

Astronomical unit (AU)	$149{,}597{,}870$ km 499 light-seconds
Earth mass	5.98×10^{24} kg
Earth radius	$6{,}378$ km
Escape velocity, Earth	11.2 km/s
Escape velocity, sun	617.7 km/s
Light-year	9.46×10^{12} km $= 63{,}240$ AU $= 0.35$ parsecs
Parsec	3.26 light-years $= 206{,}162$ AU $= 3.025 \times 10^{13}$ km
Solar mass	1.99×10^{30} kg
Solar radius	$696{,}000$ km

GLOSSARY OF ASTRONOMICAL AND PHYSICAL TERMS

Absolute magnitude: Apparent magnitude of a star if it were 10 parsecs (32.6 light-years) from Earth.

Absorption lines: Colors missing in a continuous spectrum, caused by their absorption by atoms.

Absorption spectrum: Dark lines that appear against the background of a continuous spectrum.

Acceleration: The rate of change of an object's velocity with time.

Accretion: The collection and sticking together of small particles.

Accretion disk: A disk of matter encircling a star or black hole, formed from matter pulled in from external sources (another star, for example) by the primary object's gravitational field.

Altitude: Angle at which an object appears above the horizon, measured along its vertical circle.

Angstrom (Å): Unit of length equal to one ten-billionth of a meter (10^{-10} m); used to measure very small wavelengths.

Angular diameter: Angle that the diameter of an object makes as measured at the eye of the observer.

Angular distance: Angle between two objects as viewed on the celestial sphere.

Antimatter: Subatomic particles possessing the same mass as "normal" subatomic particles such as electrons or protons, but having an opposite electrical charge and spin. They include antiprotons and positrons (antielectrons).

Aphelion: The point in an object's orbit around a star (such as the sun) at which it is the farthest from that star.

Apogee: The point in the orbit of an object around a planet (such as the Earth) at which it is farthest from the planet.

Apparent magnitude: A measure of the brightness of a star or other celestial object as seen from the Earth.

Arc minute ('): sixty arc seconds or 1/60 of a degree.

Arc second ("): A unit of angular measurement, equal to 1/3,600 of a degree; commonly used to measure the apparent diameter, annual parallax, or proper motion of a celestial object or location (such as the Galactic center) and the angular separation of binary stars.

Association: A loose cluster of stars that probably had a common origin in the same molecular cloud.

Astronomical unit (AU): The average distance from the sun to the Earth, equaling approximately 149,790,000 kilometers.

Astronomy: The science of the observational and theoretical study of celestial objects, the space between them, and the universe as a whole.

Astrophysics: The branch of astronomy that uses the tools and techniques of physics to study celestial objects.

Atom: The smallest particle of an element retaining the properties characteristic of that element.

Axion: A still-hypothetical subatomic particle that may constitute much of the missing mass in the universe, including the dark matter of the Galactic corona.

Azimuth: The angle measured along the horizon from the north point to the vertical circle that passes through a given object.

Balmer lines: A series of spectral lines in the spectrum of atomic hydrogen, produced by the jumping of electrons up from or down to the second energy level in a hydrogen atom.

Barred spiral: A spiral galaxy whose nucleus has an elongated or barred shape. The spiral arms extend from the ends of the bar. The Milky Way Galaxy may have a slightly barred nucleus.

Binary star: A double-star system in which the two stars revolve around a common point lying between them.

Blackbody: A theoretical object that perfectly absorbs and reradiates all energy that falls upon it.

Blackbody spectrum: The continuous spectrum emitted by a blackbody.

Black dwarf: The final remains of a white dwarf star after all its thermal energy has been emitted.

Black hole: An object that has collapsed under the influence of gravity and whose gravitational field is now so strong that nothing—not even light—can escape from it.

Blueshift: The decrease in the wavelength and increase in frequency of radiation emitted by an object that is approaching an observer, caused by the Doppler shift.

Bohr atom: A model of the atom devised by physicist Niels Bohr; depicts the electrons as if they were in orbit around the atom's nucleus.

Bolometric magnitude: The measure of the total radiation of a star as received above the Earth's atmosphere; measured throughout the entire electromagnetic spectrum.

Bright-line spectrum: An array of colorful lines against a dark background; produced by an excited, low-pressure gas.

Brightness: Measure of the actual luminosity of an object.

Celestial equator: Projection of the Earth's equator onto the sky.

Celestial meridian: A great circle that passes through the celestial poles and the zenith of the observer.

Celestial poles: Extensions of the Earth's axis of rotation to points on the celestial sphere about which the sky appears to rotate daily.

Celestial sphere: An apparent sphere of immense radius,

centered on the observer, upon which are located the stars, planets, and other celestial objects.

Cepheid variables: A class of variable stars that pulsate with a period of from one to seventy days and thus vary in their brightness. There are two classes of Cepheids. They are named for the star Delta Cephei, discovered in 1780 and the first of the class to be studied.

Chandrasekhar limit: The maximum amount of mass for a white dwarf star; beyond this limit a star collapses gravitationally to form a neutron star or, for very large masses, a black hole.

Cluster variables: A large class of variable stars that have a period of less than one day; usually found in globular clusters.

Color index: The difference in the magnitudes of an object as measured at two different wavelengths; a measure of a star's color and therefore its temperature.

Conjunction: The lining up of two celestial bodies so that they appear closest to each other in the sky.

Constellation: A group of stars that appear to the eye of observers from the same culture to have the shape of some object, animal, or person. Today a constellation includes the region of the celestial sphere that surrounds the configuration.

Continuous spectrum: The continuous band of color produced by a heated solid, liquid, or gas under pressure.

Continuum arc: A structure of arcing filaments near the Galactic center.

Convex lens: A lens in which one or both surfaces are curved like the surface of a sphere (a convex curvature), so that it is thicker in the middle than at the edges. Such a lens focuses light passing through it onto one point.

Copernican system: A system of planets revolving around a central sun (heliocentric).

Cosmic rays: Charged atomic particles moving through space with very high velocities.

Cosmology: The study of the origin, evolution, organization, and structure of the universe.

Cosmos: The universe considered as an orderly and self-inclusive system.

Cygnus arm: Part of one of the Galaxy's spiral arms, near or in which the sun is located.

Dark matter: The matter that makes up more than 90 percent of the Galaxy—and the entire universe—and is totally undetectable except for its gravitational effects; the "missing mass" of the Galaxy and the universe. Nearly all of the Galaxy's corona is probably made of dark matter.

Dark nebula: A cloud of dust or gas that blocks the light of stars lying behind it. Also called an absorption nebula.

Declination: The smallest angle between a celestial object and the equator; a coordinate used with right ascension in the equatorial coordinate system. The equivalent of latitude.

Deferent: In the Ptolemaic system, a large circle usually centered on the Earth, along which the center of a planet's epicycle moves.

Deflection of starlight: The bending of light by gravity as it passes by an extremely massive object.

Density: The amount of mass per volume of space, usually measured in grams per cubic centimeter.

Density waves: Waves (analogous to sound waves) that pass through interstellar matter, causing it to be compressed and triggering the formation of new stars.

Diffraction: The absorption and reemission of light as it passes by an object or through a very small opening.

Diffraction grating: A set of finely ruled lines that diffract light and produce a spectrum.

Diffuse nebula: An irregularly shaped bright nebula, usually an H_{II} region of ionized hydrogen, with an emission spectrum.

Diffusion of light: The scattering of light from an irregular surface.

Direct motion: The typical eastward motion of a planet in the night sky of Earth, seen against the background of the stars.

Dispersion: The separation of white light into its component colors by means of diffraction or refraction.

Doppler shift: The change in the observed wavelength of light or sound, caused by the motion of the observer, the source, or both.

Double star: A star system composed of two stars, each influenced by the other's gravitation.

Dwarf star: A main-sequence star of relatively low luminosity and mass; the sun is considered a dwarf star.

Eccentric: A point within a circle that is off center; commonly used in the Ptolemaic system.

Eccentricity: The measure of the degree to which an ellipse is elongated.

Eclipse: The partial or total darkening of an object by another object passing in front of it.

Eclipsing binary: A binary star system in which the plane of the stars' mutual revolution lies almost directly along the observer's line of sight, causing one star to periodically eclipse the other.

Ecliptic: The plane of the Earth's orbit projected upon the sky, and thus the apparent path of the sun on the celestial sphere.

Electromagnetic radiation: A flow or field of energy produced when an electrically charged object (such as an electron) is accelerated. Can be equally considered a wave motion or a stream of particles.

Electromagnetic spectrum: The full range of frequencies or wavelengths of electromagnetic radiation, including radio waves, infrared radiation, visible light, ultraviolet light, x-rays, and gamma rays.

Electron: A stable elementary subatomic particle carrying a negative electrical charge and having a mass of about 9.1×10^{-28} gram. The electron is a constituent of all atoms, acting as if it circles around the nucleus or exists in different energy levels or shells outside the nucleus, but also able to exist independently of atoms.

Element: Any of more than one hundred fundamental substances that cannot be reduced to simpler forms by any chemical process.

Elementary particle: A subatomic particle that is not itself composed of any other, smaller constituent. Electrons and quarks are considered elementary particles.

Ellipse: A closed and elongated curve having two axes at right angles to each other—the longer called the major axis, and the shorter the minor axis—and two foci lying on the major axis at equal distances from its central point. Orbiting bodies move along an ellipse that has the primary body at one focus.

Elliptical galaxy: A galaxy whose visible shape is that of an ellipse.

Elongation: The apparent angle between the sun and a specified celestial object.

Emission line: A bright spectral line produced by the downward transition of an electron.

Emission nebula: A nebula or cloud of gas that produces its own light, caused by the excitation of its constituent atoms.

Emission spectrum: A spectrum of bright lines produced by an excited, low-pressure gas.

Energy: The capacity to do work.

Energy level (in an atom): The various possible energies an electron may have if excited by some external source.

Epicycle: In the Ptolemaic system, a small circle along which a planet moves, and whose center moves along a deferent, thus explaining the retrograde movement of a planet.

Equant: In the Ptolemaic system, an eccentric in which the center of the circle is not the center of a celestial body's uniform motion.

Equator: A great circle on the Earth that lies halfway between the North and South poles.

Equatorial coordinate system: An astronomical coordinate system in which the fundamental reference circle is the celestial equator, the zero point is the equinox, and coordinates are declination and right ascension. The most commonly used astronomical coordinate system.

Escape velocity: The minimum velocity required for a moving body to overcome the gravitational pull of another, more massive body and never return to it.

Euclidean geometry: Geometry in which parallel lines never meet and the sum of the angles of a triangle always add up to 180 degrees; flat geometry.

Event horizon: The "boundary" of a black hole; the "surface" at which a collapsing massive body's escape velocity exceeds the speed of light, thus preventing anything from escaping.

Excitation: The process by which the electrons of an atom are imparted greater energy than they normally possess.

Expanding arm: A segment of a spiral arm or armlike structure; lies about one thousand light-years from the Galactic center and appears to be expanding away from it.

Extra-galactic: Outside of or beyond the Milky Way Galaxy.

Focal length: The distance from the center of a lens or a mirror to the prime focus of a telescope.

Focus: The point at which converging beams of light meet.

Forbidden lines: Spectral lines that cannot normally be created in an Earth-bound laboratory but that may appear under certain conditions in astronomical spectra.

Force: A physical influence that can alter a body's state of rest or of uniform motion.

Frame of reference: A set of coordinates to which the position or motion of an object in a system may be referred, and by which it may be measured.

Fraunhofer lines: Absorption lines in the spectrum of the sun or a star.

Frequency: The number of waves that pass a given point in one second.

G star: A star of spectral class G, yellow in color with a surface temperature of 5,000–6,000 K; the sun is a G star.

Galactic center: The innermost region of a galaxy; the point around which a galaxy rotates. In the Milky Way Galaxy, this point is located in the direction of the constellation Sagittarius, specifically at or very near the radio object known as Sgr A*.

Galactic cluster: An open cluster of stars found in the Galaxy's spiral arms.

Galactic corona: The outermost sphere of matter surrounding the Galaxy; thought by most astronomers to be composed mostly of some kind of "dark matter."

Galactic disk: The flattened portion of the Milky Way Galaxy, including its spiral arms.

Galactic equator: A great circle on the celestial sphere that marks the plane of the Milky Way Galaxy.

Galactic halo: An intermediate region of a galaxy, roughly spherical, lying between the outermost galactic corona and (for a spiral galaxy) the galactic disk; populated mainly by globular clusters, RR Lyrae variables, and clouds of hydrogen gas.

Galactic latitude: Angular distance north or south of the Galactic equator.

Galactic longitude: Angular distance along the Galactic equator from a zero point in the direction of the Galactic center.

Galactic plane: The plane that passes most nearly through the central plane of the Galaxy's spiral disk. The Galactic plane is inclined to the celestial equator by an angle of about 62°.

Galactic poles: The points on the celestial sphere that are 90° north and south of the Galactic equator. The north Galactic pole is located in the constellation Coma Berenices; the south Galactic pole is in the constellation Sculptor.

Galactic rotation curve: A curve plotted on a diagram showing the relationship between an object's distance from the center of the Galaxy and its velocity. Galactic rotation curves for objects that are very distant from the Galactic center revealed the existence of the dark matter Galactic corona.

Galactic supercluster: A grouping of galaxies that is made up of hundreds and even thousands of individual galaxies.

Galaxy: A vast assembly of stars, gas and dust, bound

together by and organized by its own gravity, and containing anywhere from a few million to several trillion stars. When capitalized, the word refers to our own galaxy, the Milky Way Galaxy.

Gamma rays: Very high energy electromagnetic radiation having wavelengths less than those of x-rays.

Giant molecular clouds (GMCs): Vast clouds of cool interstellar gas and dust that are the main locations of star formation in the Galaxy; composed mainly of molecular hydrogen and about 1 percent interstellar dust.

Giant star: A highly luminous star with a very large radius, usually grouped in luminosity classes II and III and having an absolute magnitude brighter than 0. Giant stars are a late phase of stellar evolution. Two typical giants are Arcturus and Capella.

Globular cluster: A large but compact spherical system of older Population II stars, containing anywhere from several tens of thousands to a million stars, which follow elliptical orbits around the center of a galaxy; usually but not always found in a galaxy's halo.

Globule: A small, dense, dark cloud of gas that is in the process of becoming a star; found in molecular clouds.

Gravitation: The mutual attractive force that all material bodies exert on one another; the most familiar of the four basic forces of nature, first expressed mathematically by Newton.

Gravitational collapse: The unopposed collapse of a celestial body or system caused by its own gravitational pull. In astronomy, gravitational collapse usually refers to the collapse of the core of a star at the end of its evolutionary cycle, resulting in the creation of a white dwarf, a neutron star, or a black hole.

Gravitational constant: A universal constant used in Newton's law of gravitation, which is the force of attraction between two bodies of unit mass separated by a unit of distance.

Gravitational energy: Energy released by the partial or total gravitational collapse of a celestial body or system.

Gravitational redshift: The redshift of spectral lines that occurs when any form of electromagnetic radiation, including visible light, is escaping from an extremely massive object. The radiation loses energy in the process, with a subsequent decrease in its frequency and increase (reddening) of its wavelength.

Gravity (of Earth): The force of attraction exerted by the Earth upon any given object on or near it.

Great circle: The largest circle that can be drawn upon a sphere, thus dividing it into two equal parts.

Great Rift: A chainlike complex of large dark nebulas that obscure the visible light coming from a band of the Milky Way stretching from Cygnus to Sagittarius.

H_I region: A region of space containing neutral hydrogen, that is, normal hydrogen atoms.

H_{II} region: A region of space containing ionized hydrogen gas, that is, hydrogen atoms from which the electrons have been removed by some energetic process.

Harmonic law: Kepler's third law of planetary motion, which states that the square of a planet's orbital period is directly proportional to the cube of its average distance from the sun.

Heliocentric system: An astronomical system in which the sun is at the center of the solar system.

Hertz (Hz): The basic unit of frequency, defined as the frequency of any periodic phenomenon that has a period of one second. Electromagnetic radiation has a range of frequencies running from about 3,000 Hz (3 kHz) to about 1×10^{22} Hz (1×10^{16} MHz).

Hertzsprung-Russell (H-R) diagram: A graph that plots a group of stars in order to show the relationship between their absolute magnitude or luminosity and their temperature or spectral class. The relationship was discovered independently in 1913 by Ejnar Hertzsprung and Henry Norris Russell.

Horizontal (or horizon) coordinate system: A coordinate system in which the fundamental reference circle is the astronomical horizon, and the zero point is the north point; the coordinates are altitude (angle above the horizon) and azimuth (angle relative to the north point).

Hubble classification: A classification scheme for galaxies first proposed by astronomer Edwin Hubble in 1925; groups galaxies into the three general categories of elliptical, spiral, and barred spiral, with each category further divided into several subcategories denoted by either a number (such as E2 for an elliptical galaxy) or letter (such as SBa for a type of barred spiral galaxy).

Hydrogen (H): The simplest of all elements, consisting of a proton orbited by an electron. Hydrogen is the most abundant element in the universe—about 94 percent by number of atoms, and 73 percent by total mass. Neutral hydrogen (or H_I) occurs throughout interstellar space and can be detected by its emission of radio waves at a wavelength of twenty-one centimeters. At low temperatures and high enough densities, pairs of neutral hydrogen atoms can form hydrogen molecules, which in turn are found in molecular clouds. Ionized hydrogen (H_{II}) results when neutral hydrogen (H_I) atoms are stripped of their electrons.

Hydrogen spectrum: The emission and absorption lines found in the spectrum of atomic or neutral hydrogen, including the Balmer, Lyman, and Paschen series of spectral lines.

Hydroxyl radical: The molecule OH, made of an atom each of hydrogen and oxygen bound together; the first interstellar molecule to be detected (in 1963) in the radio part of the spectrum.

Inertia: The property of matter that resists any change in its velocity, thus causing a body to remain either in a state of rest or in uniform motion in a straight line.

Inertial frame: A frame of reference in which an isolated particle is not subject to acceleration, including from a gravitational field.

Infrared astronomy: The branch of astronomy that studies celestial bodies by means of the infrared radiation they emit.

Infrared radiation: Electromagnetic radiation that has a wavelength slightly longer than that of visible red light.

Infrared sources: Celestial objects or locations that are strong emitters of infrared radiation. The Galactic center includes many strong infrared sources.

Intensity: The brightness of a light source.

Interstellar absorption (or extinction): The reduction in brightness of light from stars, caused by the absorption and scattering of the light by interstellar dust.

Interstellar dust: Microscopic dustlike grains that exist in the space between the stars; thought to be composed mainly of carbon, iron, and silicate materials, often with a mantle of ices. Found in molecular clouds and dark nebulas.

Interstellar gas: The diffuse gas that is found in the space between the stars.

Interstellar lines: Absorption spectral lines caused by absorption due to interstellar gas; seen against the spectrum of a star.

Interstellar matter: The gas and dust found in the space between the stars.

Interstellar medium (ISM): The matter found between the stars of the Galaxy, mostly confined to a thin layer in the Galactic plane, including interstellar dust, H_I and H_{II} regions, and giant molecular clouds. Constitutes about 10 percent of the Galaxy's total mass.

Interstellar reddening: The reddening of starlight caused by the scattering of the bluer components of light by intervening clouds of gas and dust.

Ion: An atom that is electrically charged as the result of gaining or losing one or more electrons from its electron shells.

Ionization: The process by which an atom gains or loses one or more electrons.

Ionized gas: Gas that has been ionized and that contains free electrons; a plasma.

Irregular galaxy: A galaxy that has an irregular shape, neither spiral nor elliptical, and usually contains mostly Population I stars and much gas and dust.

Irregular variable: A variable star whose changes in brightness do not follow a periodic pattern.

IRS 16: A strong source of infrared radiation at the center of the Galaxy; possibly a supermassive black hole.

Island universes: An early name for galaxies.

Jet: A long, thin, linear feature of bright emission seen coming out of some compact object, such as a galaxy. One example is the jet that Yoshiaki Sofue saw emerging from the Galactic core.

Jovian planet: A Jupiter-like planet.

Jupiter: The largest planet in the solar system, with an equatorial diameter of about 142,800 kilometers and a mass of about 1.9×10^{27} kilograms. When spelled with a lowercase j (a jupiter), refers to still-undiscovered celestial objects with masses much less than stars and less than brown dwarfs, which may constitute some fraction of the Galaxy's "missing mass."

Kelvin (K): The SI unit of temperature; K is absolute zero, equal to $-273.15\,^\circ$C or -459.67° F. Temperatures in kelvin can be converted to Celsius temperatures by subtracting 273.15; those in Celsius can be converted to kelvin by adding the same number.

Kepler's laws: The three fundamental laws of planetary motion first formulated by Johannes Kepler: (1) The orbit of each planet is an ellipse that has the sun at one focus. (2) Each planet revolves around the sun so that the line connecting the planet and the sun sweeps out an equal area in an equal amount of time. (3) The squares of the sidereal periods of any two planets are proportional to the cubes of their mean distances from the sun.

Kerr black hole: A black hole that is spinning.

Kerr-Newman black hole: A black hole that is spinning and possesses an electrical charge.

Kilogram (kg): In the SI system of measurement, the unit of mass. Equal to a thousand grams.

Kilometer (km): A unit of length equal to a thousand meters.

Kinetic energy: The energy of motion.

Kirchhoff's laws: Three statements that describe the formation of continuous, emission, and absorption spectra.

Large Magellanic Cloud (LMC): One of the Milky Way Galaxy's satellite galaxies; an irregular galaxy found in the southern constellation Dorado.

Law of areas: Kepler's second law of planetary motion, which states that each planet revolves around the sun so that the line connecting the planet and the sun sweeps out an equal area in an equal amount of time.

Law of ellipses: Kepler's first law of planetary motion, which states that each planet's orbit is an ellipse, with the sun at one focus and the second focus empty.

Light: Electromagnetic radiation visible to the human eye.

Light-year: The distance that light, traveling at about three hundred thousand kilometers per second, travels in one year; equal to about 9.5 trillion kilometers or about 6 trillion miles.

Local Group: The cluster of galaxies, bound together by mutual gravitational attraction, that includes our own Galaxy.

Long-period variable: A variable star whose period of brightness variation exceeds one hundred days.

Lorentz contraction: A consequence of relativistic velocity predicted by the special theory of relativity, in which the length of an object traveling at a relativistic velocity appears to an observer in another inertial frame to be shrinking along the direction of its motion.

Luminosity: The measure of an object's actual brightness.

Lyman lines: Lines in the ultraviolet part of the spectrum of atomic hydrogen, caused when an electron moves up from or down to the electron's lowest energy level or "orbit."

Magnitude: A measure of a star's brightness in which larger magnitudes stand for fainter objects.

Main sequence: The main series of stars on the H-R diagram, falling along a line running from upper left to lower right on the diagram.

Main sequence star: A star that falls on the main sequence of an H-R diagram.

Mass: The measure of the amount of matter in an object. The measurement of an object's resistance to change in motion is called inertial mass; the measurement of an object's gravitational field strength is called gravitational mass. The two types of masses appear to be equivalent.

Mass loss: The loss of mass by a star during its evolution; usually refers to the loss of mass early in a star's life by means of the T Tauri wind or to the loss of mass late in stellar evolution through a stellar wind, the ejection of a shell of gas, or a supernova explosion.

Messier catalog: A catalog of celestial objects compiled by Charles Messier in 1787.

Microwaves: Very high frequency radio waves, in the range of 3×10^{11} to 3×10^8 hertz.

Milky Way: The diffuse band of light that encircles the night sky, looking to the naked eye not unlike a river of spilled milk; actually composed of millions of distant stars and representing the flattened disk of the Galaxy as seen from the inside.

Missing mass: The mass in a galaxy or cluster of galaxies, or in the entire universe, that reveals its existence only by its gravitational effects on visible mass. Missing mass is therefore made of some kind of "dark" or invisible matter, matter that is not detectable from any known form of electromagnetic radiation. More than 90 percent of the Galaxy's mass, including nearly all the mass in the Galactic corona, is missing mass.

Molecule: A combination of two or more atoms bound together by the electrical forces between their electron shells.

Momentum: An object's mass multiplied by its velocity.

Multiple star: A group of two or more stars that are held together by mutual gravitational attraction and have complex interacting orbits around one another.

Naked singularity: A singularity that is not surrounded by an event horizon and is thus "naked" to the rest of the universe; naked singularities have not (yet) been discovered.

Nebula: A cloud of dust or gas in space; from the Latin word for "cloud."

Neutrino: An uncharged subatomic particle that has little or no mass, travels at or near the speed of light, and carries away energy from certain types of nuclear reactions.

Neutron: An uncharged subatomic particle that has about the same mass as a proton and along with protons makes up the nuclei of all atoms but hydrogen.

Neutron star: A star that has undergone gravitational collapse, creating an object that has extremely high density (more than 1×10^{17} kilograms per cubic centimeter) and very small size (20–30 kilometers in diameter) and is made mostly of tightly packed neutrons; the end product of the evolution of giant stars with a mass greater than 1.4 times that of the sun (the Chandrasekhar limit) and less than about 5.0 times that of the sun.

New General Catalog (NGC): A catalog of stars, star clusters, nebulas, and galaxies; succeeded and added to the star catalog created by Charles Messier.

Newtonian reflector: A reflecting telescope that uses an angled flat mirror to bring the rays of light from a distant object to a focus at the side of the telescope tube.

Newton's laws of motion: The three laws of motion formulated by Isaac Newton and published in the *Principia* in 1687: (1) Every body continues in a state of rest or uniform motion in a straight line unless that state is changed by the action on the body of an outside force. (2) The rate of change of linear momentum is proportional to the applied force and occurs in the same direction as that force. (3) Every action has an equal and opposite reaction.

Nonthermal emission (nonthermal radiation): Electromagnetic radiation, such as synchrotron emission, that is produced by the acceleration of electrons or other charged subatomic particles and does not have the spectrum of a blackbody.

North celestial pole: The point on the celestial sphere that is a direct extension of the Earth's rotational axis through the Earth's North Pole.

Nova: A star that suddenly becomes bright and then fades again; from the Latin word for "new." Novae are binary stars in which one of the pair suddenly flares as matter from the second falls upon its surface and undergoes a fusion reaction. Not to be confused with a supernova.

Nuclear fusion (thermonuclear fusion, thermonuclear reaction): The energy-producing process in stars; a process by which the nuclei of two light elements (such as hydrogen or helium) join under conditions of great temperature and pressure to create a heavier nucleus, with a concomitant change of matter to energy and the release of that energy.

Nucleus (atom): The central part of an atom, made of protons and neutrons and comprising almost the entire mass of the atom.

OB stars: Stars of the spectral class O and B, which are very large, young, and bright in the blue, violet, and ultraviolet regions of the spectrum.

Objective: The main lens or mirror of a telescope, which brings to a focus the rays of light from a distant object.

Occultation: The passing of one celestial object behind another.

Open cluster: A loosely formed cluster of stars, most often found in the Galactic disk.

Opposition: The configuration of a planet when it is directly opposite the sun as seen from Earth.

Optical telescope: A telescope that collects light from the visible part of the spectrum using lenses or mirrors, then focuses it to a focal point.

Optics: The branch of physics that deals with light and its properties.

Orbit: A closed path along which a celestial body moves as it revolves around a point in space, a point usually (although not in the Ptolemaic system) occupied by a celestial body of much greater mass and gravitational attraction.

Orbital plane: The plane in which a celestial body moves in its orbit.

Orion: A well-known constellation lying on the celestial equator, named for a hunter in Greek mythology; includes the stars Betelgeuse and Rigel and the Orion Nebula.

Orion Molecular Cloud (OMC-1): A dense cloud of neutral molecular hydrogen (an H_1 complex) lying behind and in contact with the Orion Nebula and containing a cluster of bright infrared sources that may be a group of new stars.

Orion Nebula (M42): One of the brightest and best-known emission nebulas in the sky, located in the constellation Orion about 3,700 light-years from Earth and visible to the naked eye as a faint fuzzy patch about thirty-five arc minutes wide. It is a complex region of ionized hydrogen (an H_{II} region) containing the Trapezium cluster of new stars, and is a region of active star formation.

Parallax: The apparent shift in position of an object when it is viewed from two different locations. Thus, a stellar parallax is the apparent shift in position of a star against the background of more distant (and relatively unmoving) stars caused by the motion of the Earth around the sun. Also called parallax shift.

Parsec: The distance to a star that has a parallax shift of one second of arc; equal to about 3.26 light-years.

Paschen series: A series of spectral lines found in the spectrum of hydrogen.

Peculiar galaxies: Galaxies that do not fit into the standard classification schemes for galaxies.

Penrose diagram: A two-dimensional diagram depicting four-dimensional space-time. Three-dimensional space is represented by a horizontal line, and time by a vertical line. Objects moving at light speed are represented by two light lines lying at forty-five-degree angles to the vertical and horizontal lines. Worldlines lying between the light lines and the vertical time line depict the movement through space-time of all objects in the universe.

Perfect radiator: A blackbody.

Periastron: The point in an orbit of a star in a binary system that is closest to the second star.

Perigalacticon: The point in a star's orbit around the Galaxy that is nearest the Galactic center.

Perigee: The point in the orbit of a satellite around the Earth that is closest to the Earth.

Perihelion: The point in the orbit of a celestial body around the sun that is closest to the sun.

Period: The interval of time needed for one revolution, one rotation, or one cycle.

Period-luminosity relation: The relationship between the luminosity of a Cepheid variable star and its period light variation. Discovered by Henrietta Leavitt in 1912.

Perturbation: Any gravitational disturbance that causes a body to deviate from its normal orbital path.

Photoelectric effect: The emission of electrons from the surface of a substance exposed to electromagnetic radiation above a specific frequency.

Photometry: The branch of astronomy that measures the apparent brightness of celestial objects.

Photon: A unit or quantum of electromagnetic energy; a "particle" of light.

Planck's constant (h): A dimensionless number invented by Max Planck; relates the energy carried by a photon to the photon's wavelength.

Planet: A celestial body that orbits the sun or another star and shines only by the light it reflects.

Planetary nebula: A slowly expanding spherical shell of gas ejected by a dying red giant star late in its evolution.

Plasma: A state of matter made of freely moving electrons and ions; an ionized gas. The other three states of matter are solids, liquids, and gases.

Pleiades: An open cluster of stars about 390 light-years from Earth in the constellation Taurus, still embedded in some of its original gas and dust cloud, which glows from reflected starlight. Also known as the Seven Sisters and M45.

Population I stars: A class of stars found mainly in the disk and arms of a spiral galaxy. These stars are relatively young, metal-rich, and highly luminous.

Population II stars: A class of stars found mainly in elliptical galaxies and in the centers and halos of spiral galaxies. These stars are older, redder, and more metal-poor than Population I stars.

Population III stars: A hypothetical population of stars that some astronomers believe might have existed before the formation of the Galaxy.

Positron: The antimatter equivalent of an electron; a sub-atomic particle with the mass of an electron but an opposite (that is, positive) electrical charge and opposite spin.

Potential energy: The energy possessed by an object because of its position; equal to the work done by the body as it changes from its given state or position to some standard state or position.

Prime focus: The point at which the objective of a telescope brings the rays of light to a focus without the use of any secondary mirror or lens.

Principle of equivalence: The statement that one cannot distinguish between different kinds of acceleration and/or that gravitational and inertial mass are equivalent; a key idea in Einstein's general theory of relativity.

Prism: A triangular wedge of glass or some other transparent material that disperses white light into its component colors.

Proper motion: The rate at which the position of a star in the sky changes; measured in arc seconds per year.

Proton: A subatomic particle that has a positive electrical charge equal in strength to the negative charge of an electron and a mass about 1,836 times that of an electron; found in the nuclei of all atoms.

Protostar: A collapsing mass of gas and dust that is in the process of becoming a star; an early stage in stellar evolution.

Pulsar: A rapidly spinning neutron star emitting a beam of radiation that is detected as "pulses" by Earth observers.

Quantum: The minimum amount by which certain properties of a system can change, with the result that the value of the property can change only in discrete steps and not continuously. For example, the photon is a quantum of electromagnetic energy.

Quantum mechanics: The branch of physics founded by

Max Planck in 1900 and brought to fruition by Niels Bohr in the 1920s; studies the structure and interaction of atoms and their components as properties that change only in discrete increments.

Quantum physics: Another commonly used name for quantum mechanics.

Quasar: A compact extra-Galactic object that looks like a faint point of light, emits more energy than hundreds of galaxies, and often has a strong redshift, indicating a great distance from the Milky Way. Many astronomers now believe from some observational evidence that quasars are the violently active cores of distant, active galaxies—galactic nuclei more active than those of Seyfert galaxies.

Radial velocity: The portion of an object's velocity that is measured along the observer's line of sight.

Radio astronomy: The branch of astronomy that studies celestial objects by means of their emitted radio waves.

Radio source: A celestial object that can be detected with a radio telescope. For example, the Galactic center is a strong radio source.

Radio telescope: A telescope that collects radio waves instead of visible light, and focuses it on a focal point; often but not always a dish-shaped antenna. Data collected by a radio telescope can be constructed using computer techniques into a "picture" resembling that made by an optical telescope.

Radio waves: Electromagnetic radiation whose frequencies lie in the radio part of the spectrum, from about 20 kilohertz to 300 gigahertz.

Red giant: A very large star with a surface temperature of 2,000–3,000 K and a diameter ten to one hundred times that of the sun; one of the final phases in the evolution of a normal star.

Redshift: The displacement of spectral lines toward the redder end of the spectrum, usually caused either by the object moving away from the observer (the Doppler effect) or by gravitational redshift.

Reflecting telescope: An optical telescope that uses a

mirror to bring the rays of light to a primary focus; invented by Isaac Newton.

Reflection nebula: A nebula that is visible because it reflects the light of nearby stars.

Reflector: Reflecting telescope.

Refracting telescope: A telescope that uses lenses to bring rays of light to a primary focus. The first "spyglass," invented by Hans Lippershey in 1608 and copied by Galileo in 1609, was a refractor.

Refractor: Refracting telescope.

Reissner-Nordström black hole: A black hole that has an electrical charge but is not spinning.

Relativistic velocity: A velocity that is close to or is a substantial fraction of the speed of light; at relativistic velocities, strange phenomena such as time dilation and Lorentz contraction occur.

Relativity, general theory: A theory formulated by Albert Einstein in 1915 to describe the relationship between space and time as affected by gravity; a theory of gravitation.

Relativity, special theory: A theory formulated by Albert Einstein in 1905 dealing with the measurement of events seen by observers moving relative to one another in different inertial frames.

Resolution: The ability of a telescope to separate objects that appear to lie very close to each other; the ability to show great detail.

Rest mass: The mass of an object when it is at rest. The mass of an object increases as its velocity increases, becoming noticeable at velocities close to that of light.

Retrograde motion: The apparent westward motion of a planet as seen against the background of fixed stars.

Revolution: The movement of a body around a given point in space.

Right ascension: A coordinate used with declination in the equatorial coordinate system, measuring the angular distance from a celestial body eastward along the celestial equator from the zero hour circle to the intersection of the hour circle passing through the body. Generally expressed in hours, minutes, and seconds from 0 to 24, with one hour equaling fifteen degrees of arc. The equivalent of longitude.

Rotation: The spinning of a body on its own axis.

RR Lyrae variable: A class of variable stars having a period of less than one day; often found in globular clusters and as isolated stars in the Galactic halo.

Sagittarius: A constellation (the Archer) of the zodiac in the Southern Hemisphere, part of which lies in the Milky Way; includes such astronomical objects as the Trifid Nebula, the Omega Nebula, and several globular clusters. The Galactic center lies in the direction of Sagittarius.

Sagittarius A (Sgr A): The brightest of several radio sources at the Galactic center; includes two components, Sagittarius A East and Sagittarius A West.

Sagittarius A East (Sgr A East): One of two components of Sgr A, probably a supernova remnant near the Galactic center.

Sagittarius A West (Sgr A West): One of two components of Sgr A, a source of intense thermal emission from ionized gas, as well as the location of several strong sources of gamma rays, x-rays, infrared radiation, and radio waves. Two such sources, Sgr A* and IRS 16, are candidates for the location of a supermassive black hole at the Galactic center.

Sagittarius A* (Sgr A*): An intense point radio source within Sgr A West at the center of the Galaxy; possibly a supermassive black hole.

Sagittarius B (Sgr B): A giant molecular cloud (GMC) complex lying near the Galactic center, possibly part of an expanding ring of such clouds at the center.

Schwarzschild black hole: A black hole that has neither any spin nor any electrical charge; the simplest form of black hole, first predicted from Einstein's general relativity equations in 1916 by Karl Schwarzschild.

Schwarzschild radius: The size of a massive object at which its density and therefore its gravitational field are so strong that not even light can escape it; the radius of a black hole's event horizon.

Separation: The angular distance between two stars in a visual binary system.

Seyfert galaxy: A class of spiral galaxies, first discovered by Carl Seyfert in 1943, that have extremely active and luminous nuclei.

Shock wave: A compression wave produced in a liquid, gas, or plasma by a sharp, violent change in pressure, density, or temperature.

SI system: The system of measurement units most commonly used in science today. "SI" stands for *Système International.* Its basic unit of length is the meter, of mass is the kilogram, of force is the newton, and of energy is the joule; other units include the second (time), the electron volt (energy in atomic physics), and the pascal (pressure).

Sidereal day: A day as measured by the stars; the length of a day measured by the successive passages of a particular star past the observer's meridian.

Singularity: The point of zero dimension and infinite density at the center of a black hole, into which any object collapses and whose gravitational field is so strong that nothing can resist its inward pull. All laws of nature cease to function in a singularity.

Small Magellanic Cloud (SMC): The second of two satellite galaxies to the Milky Way observed by Ferdinand Magellan in 1519; a small Type I irregular galaxy (Irr I) located in the Southern Hemisphere constellation Tucanae.

Solar mass: The astronomical unit of mass, equal to the mass of the sun (1.9891×10^{30} kilograms).

Solar wind: A flow of energetic charged particles from the sun's outer regions that expands outward from the sun past the planets and eventually into interstellar space.

South celestial pole: A point on the celestial sphere marked by the extension of the Earth's rotational axis from the Earth's south rotational pole.

Space-time: The unified system of three space and one time dimensions first revealed by Einstein in his special and general relativity theories; space and time considered as one entity.

Space-time diagram: Penrose diagram.

Spectral class: The classification of a star according to the characteristics of its spectrum.

Spectral line: A wavelength of light that corresponds to an energy transition in an atom; a dark or bright line in an absorption or emission spectrum.

Spectrogram: A photograph of a spectrum.

Spectrograph: An instrument used to photograph a spectrum.

Spectroscope: An instrument used to examine the spectrum of an object, such as a star or galaxy, or of an atomic element.

Spectroscopic binary star: A binary star system whose double nature is revealed only by periodic shifting of its spectral lines.

Spectrum: The rainbow of colors produced when light is dispersed by a prism or some other instrument of refraction or diffraction. Also, the amount of energy given off by an object at every electromagnetic wavelength.

Speed: The rate at which the distance to an object changes, without regard to the direction of the object's motion.

Speed of light: The constant speed at which electromagnetic radiation travels through a vacuum, equal to about 300,000 kilometers per second or 186,000 miles per second.

Spiral arms: In a galaxy, the curved armlike structures made of stars, gas, and dust clouds that surround the nucleus, lying more or less in the galaxy's equatorial plane. Current astronomical theory holds that spiral arms are formed by the passage of density waves.

Spiral galaxy: A galaxy with a flattened, disklike shape that contains a central nucleus surrounded by series of spiral arms that appear to trail out of the nucleus.

Star: A spherical mass of gas and plasma that generates energy by means of nuclear fusion reactions in its center and radiates various wavelengths of electromagnetic radiation. All stars have a mass greater than 0.05 solar masses, since below that mass there is insufficient density and temperature in the core to trigger fusion reactions.

Starburst galaxy: A spiral galaxy in which a massive burst of star formation is taking place.

Star cloud: A region in the sky where the stars are so close together that they appear to resemble a cloud. The Milky Way is the best example of a star cloud or series of star clouds.

Star cluster: Any grouping of stars held together by mutual gravitational interaction.

Star gauges: Systematic counts of stars in different regions of the sky.

Stellar evolution: The life cycle of a star.

Stellar wind: The steady flow of plasma emitted by many stars, including the sun (called a solar wind).

Subdwarf star: A star smaller than a dwarf star and less luminous than a main sequence star of the same spectral class.

Subgiant star: A star whose luminosity and size are midway between those of a normal star and a giant star of the same spectral class.

Subluminous star: A star that is much fainter than main sequence stars. This category includes white dwarfs, subdwarfs, and high-velocity stars.

Supergiant star: A massive star of extremely high luminosity and great size.

Supermassive black hole: A black hole of any type (Schwarzschild, Kerr, Kerr-Newman, or Reissner-Nordström) that has a mass of a million or more solar masses; thought to exist at the center of many galaxies, including our own.

Supernova: The catastrophic explosion of a giant or supergiant star, briefly increasing its luminosity to greater than that of an entire galaxy. The core left behind after a supernova explosion may be a neutron star or black hole.

Supernova remnant (SNR): The stellar core left behind after a supernova explosion; usually a neutron star or black hole.

Synchrotron radiation: A type of radiation released by charged subatomic particles (such as electrons) spiraling along lines of magnetic force.

T Tauri stars: A class of variable stars with rapid and irregular pulsations and strong stellar winds, always found in groups and embedded in dense patches of gas and dust. T Tauri stars are very young, and are the youngest optically observable stage of a star's life.

T Tauri wind: A continual and extremely strong flow of matter (a stellar wind) coming from the surface of a T Tauri star at a rate that often approaches one ten-millionth of a solar mass per year.

Tangential velocity: The part of a star's spatial velocity that is perpendicular to its radial velocity, that is, crossing the observer's line of sight.

Telescope: A light-gathering instrument using lenses, mirrors, or some combination of the two in order to form an image of a distant and/or very dim object.

Temperature: A measure of the heat energy of an object.

Thermal emission: Electromagnetic radiation caused by the interaction of electrons and atoms or molecules in a hot, dense medium, usually an ionized gas; often characterized by a blackbody spectrum.

Thermal radiation: Another term for thermal emission.

Time dilation: A consequence of relativistic velocity between two observers, predicted by the special theory of relativity and subsequently verified by experiment, by which each of two observers approaching at a relativistic velocity sees the other's clock running more slowly than his or her own.

Trapezium: A cluster of four young stars lying in the Orion Nebula that excite and ionize the nebula.

Ultraviolet astronomy: The branch of astronomy that studies celestial objects by observing the ultraviolet radiation they emit.

Ultraviolet radiation: The part of the electromagnetic spectrum that includes wavelengths shorter than visible light and longer than x-rays and gamma rays.

Universe: All of space that is occupied by matter and energy.

Variable star: A star that varies in luminosity or color or both.

Velocity: The rate and direction at which distance is covered in some interval of time; a quantity denoting both the direction and speed of a body's motion.

Volume: A measurement of the amount of space occupied by an object.

Vulcan: An imaginary planet once thought to lie within the orbit of Mercury. Vulcan's existence was postulated to explain discrepancies in Mercury's orbit.

Watt: A unit of power equal to ten million ergs used in one second.

Wavelength: The distance between the crests of two successive waves.

Weight: A measure of the force exerted on one object by the gravitational pull of another.

White dwarf: An old star with less than 1.4 solar masses (the Chandrasekhar limit) that has undergone gravitational collapse; has a diameter less than 1 percent that of the sun and an absolute magnitude of +10 to +15. White dwarfs are the final phase in the evolution of low-mass stars.

White hole: The reverse of a black hole; a theoretical region where matter and energy appear that have fallen into a black hole in another universe.

Wien's law: A statement that relates the temperature of a body to the wavelength of its maximum radiation.

Wolf-Rayet stars: A class of stars that have extemely high temperatures (sometimes as much as 50,000 K) and that eject shells of gas at high velocity.

Worldline: The path of a particle through space-time; also, such a path as depicted on a space-time or Penrose diagram.

Wormhole: A theoretical connection between a Kerr or Kerr-Newman black hole in one universe and a white hole in another universe.

X-ray astronomy: The branch of astronomy that studies objects in the universe by the x-rays they emit.

X-ray sources: Sources of x-ray emission lying beyond the solar system. The Galactic center is an x-ray source.

X-ray telescope: An instrument that is carried above the Earth's atmosphere by plane, balloon, or satellite and can detect, focus, and record x-ray emissions from space.

X-rays: Electromagnetic radiation with wavelengths shorter than ultraviolet radiation and longer than gamma radiation.

Year: The time required for the Earth to make one revolution around the sun.

Zenith: The point on the celestial sphere lying directly above the observer and ninety degrees from the observer's horizon.

Zero-age main sequence: The position on the H-R diagram reached by a protostar when it "turns on" and begins producing its energy from fusion reactions.

Zodiac: A band on the celestial sphere centered upon the ecliptic, containing twelve constellations that are traditionally associated with astrology (the "signs of the zodiac").

BIBLIOGRAPHY

This bibliography lists books, articles, and scientific preprints that were used as primary and secondary source material for *Journey to the Center of Our Galaxy*. Not all the information in all these sources found its way into this book. Most of what did not, however, was useful as background information.

Of the textbooks and popular science books listed, several were valuable sources of information and background material about physics, astronomy, black holes, and the Milky Way Galaxy. They included Zeilik's *Astronomy: The Evolving Universe*, Illingworth's *The Facts on File Dictionary of Astronomy*, Greenstein's *Frozen Star*, Hodge's *Galaxies*, Nicholson's *Gravity, Black Holes and the Universe*, and Narlikar's *Violent Phenomena in the Universe*.

Excellent sources of historical information included Weaver's *The World of Physics: A Small Library of the Literature of Physics from Antiquity to the Present*, Cohen's *Revolution in Science*, Berry's *A Short History of Astronomy: From Earliest Times Through the Nineteenth Century*, and Crowe's *Theories of the World from Antiquity to the Copernican Revolution*.

The major sources of information about mythologies and cosmic worldviews were Campbell's *Historical Atlas of World Mythology, Volume 1: The Way of Animal Powers*, Bierhorst's *The Mythology of North America*, and Eliot's *Myths*.

Particularly valuable as primary sources of scientific information were *Astronomical Journal, Astrophysical Journal, Astrophysical Journal Letters, Nature*, and *Science*. In addition, *Nature* and the *Annual Review of Astronomy and Astrophysics* were excellent sources of review articles on relevant topics.

Several scientific journals and more popular magazines proved to be good secondary sources of information. They included *Astronomy, Nature, Science, Scientific American*, and *Sky & Telescope*.

Allen, David A., and Robert H. Sanders. Is the Galactic Centre Black Hole a Dwarf? *Nature*, January 16, 1986, pp. 191–94.

Alvarez, H., et al. The Rotation of the Galaxy Within the Solar Circle. *Astrophysical Journal*, January 10, 1990, pp. 495–502.

Anderson, Per H. Massive Objects in Galactic Nuclei May Be Black Holes. *Physics Today*, October 1987, p. 22.

Asimov, Isaac. *Foundation*. New York: Doubleday & Co., 1951.

Bartusiak, Marcia. Coming Home. *Discover*, September 1988, pp. 30–37.

Begley, Sharon. Where the Wild Things Are. *Newsweek*, June 13, 1988, pp. 60–65.

Benford, Gregory. *Tides of Light*. New York: Bantam Books, 1989.

Berry, Arthur. *A Short History of Astronomy: From Earliest Times Through the Nineteenth Century*. New York: Dover Publications (repr. 1898, John Murray), 1961.

Bierhorst, John. *The Mythology of North America*. New York: William Morrow & Co., 1985.

Black Heart of the Sombrero Galaxy. *Sky & Telescope*, January 1989, p. 12.

Black Holes: Coming to a Galaxy Near You. *Discover*, October 1987, p. 10.

Black-Hole Bicentennial. *Sky & Telescope*, October 1984, pp. 312–13.

Blitz, Leo, Michel Fich, and Shrinivas Kulkarni. The New Milky Way. *Science*, June 17, 1983, pp. 1233–40.

Blumenthal, George R., et al. Formation of Galaxies and Large-Scale Structures with Cold Dark Matter. *Nature*, October 11, 1984, pp. 517–25.

Bok, Bart. A Bigger and Better Milky Way Galaxy. *Astronomy*, January 1984, pp. 6–22.

Bronfman, L., et al. CO Survey of the Southern Milky Way: The Mean Radial Distribution of Molecular Clouds Within the Solar Circle. *Astrophysical Journal*, January 1, 1988, pp. 248–66.

Brown, Robert L., and Harvey S. Liszt. Sagittarius A and Its Environment. *Annual Review of Astronomy and Astrophysics*, 1984, pp. 223–65.

Budiansky, Stephen. Gazing Back 17 Billion Years. *U.S. News & World Report*, January 25, 1988, pp. 63–64.

Burrows, Adam. The Birth of Neutron Stars and Black Holes. *Physics Today*, September 1987, pp. 28–37.

Campbell, Joseph. *Historical Atlas of World Mythology, Volume 1: The Way of Animal Powers*. New York: Harper & Row, 1983.

Carney, Bruce W., et al. Is the Galactic Halo Related to the Disk or to the Bulge? *ESO/CTIO Workshop on Bulges of Galaxies, Preprint*, 1990.

———. A Survey of Proper-Motion Stars, VIII: On the Galaxy's Third Population. *Astronomical Journal*, February 1989, pp. 423+.

———. A Survey of Proper-Motion Stars, IX: The Galactic Halo's Metallicity Gradient. *Astronomical Journal*, January 1990, pp. 201+.

———. A Survey of Proper-Motion Stars, X: The Early Evolution of the

Galaxy's Halo. *Astronomical Journal*, February 1990, pp. 572+.

Catchpole, Robin M. A Window on Our Galaxy's Core. *Sky & Telescope*, February 1988, pp. 154–55.

Chaisson, Eric. Journey to the Center of the Galaxy. *Astronomy*, August 1980, pp. 8–22.

Cohen, J. Bernard. *Revolution in Science*. Cambridge, MA: Harvard University Press, 1985.

Cohen, R. S., et al. Molecular Clouds in the Carina Arm. *Astrophysical Journal*, March 1, 1985, pp. L15–L20.

Cooke, Donald A. *The Life and Death of Stars*. New York: Crown Publishers, 1985.

Crawford, M. K., et al. Mass Distribution in the Galactic Centre. *Nature*, June 6, 1985, pp. 467–70.

Crowe, Michael J. *Theories of the World from Antiquity to the Copernican Revolution*. New York: Dover Publications, Inc., 1990.

Dame, T. D., et al. A Composite CO Survey of the Entire Milky Way. *Astrophysical Journal*, November 15, 1987, pp. 706–720.

Dame, Thomas M. The Molecular Milky Way. *Sky & Telescope*, July 1988, pp. 22–27.

Danly, Laura. Kinematics of Milky Way Halo Gas, I: Observations of Low-Ionization Species. *Astrophysical Journal*, July 15, 1989, pp. 785–806.

Danly, Laura, and Chris Blades. Ultraviolet Observations of Halo Clouds. *Space Telescope Science Institute, Preprint No. 377*, September 1989.

Dixon, Robert T. *Dynamic Astronomy*. Englewood Cliffs, NJ: Prentice-Hall, 1984.

Djorgovski, S., and C. Sosin. The Warp of the Galactic Stellar Disk Detected in IRAS Source Counts. *Astrophysical Journal Letters*, June 1, 1989, pp. L13–L16.

Eliot, Alexander. *Myths*. New York: McGraw-Hill, 1976.

Feeding the Hole. *Discover*, June 1989, p 14.

Feeding the Monster. *Sky & Telescope*, December 1986, pp. 574–75.

Freeman, K. C. The Galactic Spheroid and Old Disk. *Annual Review of Astronomy and Astrophysics*, 1987, pp. 603–632.

Frogel, Jay A. The Galactic Nuclear Bulge and the Stellar Content of Spheroidal Systems. *Annual Review of Astronomy and Astrophysics*, 1988, pp. 51–92.

Furley, David J. *The Greek Cosmologists*, Vol. 1. New York: Cambridge University Press, 1987.

The Galactic Center. *Scientific American*, March 1989, p. 20D.

Gamma-Ray Confusion at the Galactic Center. *Sky & Telescope*, June 1989, pp. 584–85.

Gatley, Ian, et al. A Large Near-Infrared Image of the Galactic Center. In Mark Morris, ed., *The Center of the Galaxy* (for the International Astronomical Union). Norwell, MA: Kulwer Academic Publishers, Inc., 1989, pp. 361–64.

Genzel, R., and C. H. Townes. Physical Conditions, Dynamics, and Mass Distribution in the Center of the Galaxy. *Annual Review of Astronomy and Astrophysics*, 1987, pp. 377–423.

Gilmore, Gerry, and Bob Carswell. *The Galaxy* (NATO ASI Series, Series C, Mathematical and Physical Sciences, Vol. 207). Dordrecht, The Netherlands: D. Reidel Publishing Co., 1987.

Glass, I. S., et al. J, H and K Maps of the Galactic Centre Region, II: Qualitative Aspects of the Interstellar Absorption. *Monthly Notes of the Royal Astronomical Society*, July 1987, pp. 373–79.

Gabrelsky, D. A., et al. Molecular Clouds in the Carina Arm: Large-Scale Properties of Molecular Gas and Comparison with H_I. *Astrophysical Journal*, April 1, 1987, pp. 122–41.

———. Molecular Clouds in the Carina Arm: The Largest Objects; Associated Regions of Star Formation, and the Carina Arm in the Galaxy. *Astrophysical Journal*, August 1, 1988, pp 181–96.

The Great Galactic Ring. *Sky & Telescope*, February 1986, p. 147.

Greenstein, George. *Frozen Star*. New York: Freundlich Books, 1983.

Grun, Bernard. *The Timetables of History: A Horizontal Linkage of People and Events*. New York: Simon and Schuster, 1979.

Hartmann, William K. *Cycles of Fire: Stars, Galaxies and the Wonder of Deep Space*. New York: Workman Publishing Co., 1987.

Haud, U. Our Galaxy—A Polar Ring Galaxy? *Astronomy and Astrophysics*, June 1988, pp. 1–3.

Heart of Darkness. *Scientific American*, August 1985, p. 66.

Hodge, Paul. *Galaxies*. Cambridge, MA: Harvard University Press, 1986.

Horgan, John. Cosmic Forgery: Can Black Holes, Like Supernovas, Make Elements? *Scientific American*, April 1988, pp. 20, 22.

———. Cosmic News: 1987A, the Milky Way's Core, Dark Matter Lenses, and More. *Scientific American*, March 1989, p. 21A.

How Many Black Holes in the Milky Way? *Sky & Telescope*, December 1983, p. 502.

Illingworth, Valerie. *The Facts on File Dictionary of Astronomy*. New York: Facts on File Publications, 1985.

Is the Galactic Center Exploding? *Sky & Telescope*, January 1988, p. 10.

Kahan, Gerald. $E = mc^2$: *Picture Book of Relativity*. Blue Ridge Summit, PA: TAB Books, Inc., 1983.

Kanipe, Jeff. Quest for the Most Distant Objects in the Universe. *Astronomy*, June 1988, pp. 20–27.

Kassim, N. E., T. N. LaRosa, and W. C. Erickson. Steep-Spectrum Radio Lobes Near the Galactic Centre. *Nature*, August 7, 1986, pp. 522–24.

Killian, Anita. Galactic Center Update. *Sky & Telescope*, March 1986, p. 255.

Koyre, Alexandre. *From the Closed World to the Infinite Universe*. Baltimore: Johns Hopkins University Press, 1957.

Krasheninnikova, Y., et al. Configuration of Large-Scale Magnetic Fields in Spiral Galaxies. *Astronomy and Astrophysics*, April 1989, pp. 19–28.

Kraus, John. The Center of Our Galaxy. *Sky & Telescope*, January 1983, pp. 30–31.

Kühn, Ludwig. *The Milky Way: The Structure and Development of Our Star System*. New York: John Wiley & Sons, 1982.

Large Magellanic Cloud: More than Meets the Eye. *Sky & Telescope*, July 1988, p. 11.

Lesch, H., et al. The Formation of Molecular Rings in Galactic Central Regions. *Monthly Notes of the Royal Astronomical Society*, January 15, 1990, pp. 194–99.

Leventhal, M., et al. Reappearance of the Annihilation Line Source at the Galactic Centre. *Nature*, May 4, 1989, pp. 36–38.

Lewis, J. R., and K. C. Freeman. Kinematics and Chemical Properties of the Old Disk of the Galaxy. *Astronomical Journal*, January 1989, pp. 139–62.

Lo, K. Y. The Galactic Center: Is It a Massive Black Hole? *Science*, September 26, 1986, pp. 1394–1402.

Lo, K. Y., and M. J. Claussen. High-Resolution Observations of Ionized Gas in Central 3 Parsecs of the Galaxy: Possible Evidence for Infall. *Nature*, December 15, 1983, pp. 647–51.

Lo, K. Y., et al. On the Size of the Galactic Centre Compact Radio Source: Diameter < 20 AU. *Nature*, May 9, 1985, pp. 124–26.

Looking Down on the Milky Way. *Sky & Telescope*, August 1989, p. 127.

Lu, J. F. Accretion Disk–Driven Jet Precession in Active Galactic Nuclei. *Astronomy and Astrophysics*, 1990, vol. 229, pp. 424–26.

Lu, J. F., and S. Pineault. Jets with Angular Momentum from the Vicinity of Kerr Black Holes. *Astronomy and Astrophysics*, 1990, vol. 229, pp. 416–23.

Luminet, J.-P., and B. Barbuy. Chemical Enrichment by Tidally Disrupted Stars Near a Black Hole in the Galactic Center. *Astronomical Journal*, March 1990, pp. 838–42.

Maran, Stephen P. A Black Hole for Sure? Astronomers Have the Best Evidence Yet. *Natural History*, February 1985, pp. 26–28.

McClintock, Jeffrey. Do Black Holes Exist? *Sky & Telescope*, January 1988, pp. 28–33.

Melott, Adrian L. Our Cosmic Horizons, Part 4: Re-Creating the Universe. *Astronomy*, May 1988, pp. 42–47.

Mikkola, S., and M. J. Valtonen. The Slingshot Ejections in Merging Galaxies. *Astrophysical Journal*, January 10, 1990, pp. 412–20.

The Milky Way's Jet. *Sky & Telescope*, August 1989, pp. 127–28.

The Milky Way's Massive Halo. *Sky & Telescope*, December 1983, pp. 500–501.

Mirabel, Igor F. The Magellanic Stream and Other Hydrogen Remnants of Strong Tidal Disruption of the Magellanic Clouds. *Astrophysical Journal*, November 15, 1981, pp. 528–33.

Molecular Clouds in the Inner Galaxy. *Sky & Telescope*, November 1986, p. 466.

Morris, Mark, editor. *The Center of the Galaxy: Proceedings of the 136th Symposium of the International Astronomical Union, Held in Los Angeles, USA.* Dordrecht, the Netherlands: Kluwer Academic Publishers, 1989.

Morris, Mark, and Farhad Yusef-Zadeh. The Center of the Galaxy: A Summary of IAU Symposium No. 136. *Nuclear Physics B (Proceedings Supplement)*, 1989, vol. 10B, pp. 59–66.

Moskowitz, Bruce. The Search for Dark Matter in the Laboratory. *New Scientist*, April 15, 1989, pp. 39–42.

Motz, Lloyd, and Anneta Duveen. *Essentials of Astronomy.* New York: Columbia University Press, 1977.

Nagata, Tetsuya, et al. AFGL 2004: An Infrared Quintuplet Near the Galactic Center. *Astrophysical Journal*, March 1, 1990, pp. 83–88.

Narlikar, Jayant V. *Violent Phenomena in the Universe.* Oxford: Oxford University Press, 1982.

New Black Hole Candidate. *Sky & Telescope*, October 1986, p. 4332–33.

Nicholson, Iain. *Gravity, Black Holes and the Universe.* New York: John Wiley & Sons, 1981.

Niven, Larry. *Neutron Star.* New York: Ballantine Books, 1968.

Noh, Hye-Rim, and John Scalo. History of the Milky Way Star Formation Rate from the White Dwarf Luminosity Function. *Astrophysical Journal*, April 1, 1990, pp. 605–614.

Of Whirlpools, Warps, Bubbles, and Jets. *Sky & Telescope*, January 1988, pp. 9–10.

Okuda, Haruyuki, et al. An Infrared Quintuplet Near the Galactic Center. *Astrophysical Journal*, March 1, 1990, pp. 89–97.

Our Active Galactic Nucleus. *Sky & Telescope*, December 1986, p. 577.

Pagels, Heinz R. Before the Big Bang. *Natural History*, April 1983, pp. 22–26.

Parker, Barry. In and Around Black Holes. *Astronomy*, October 1986, pp. 6–15.

Peters, Philip C. Black Hole. *McGraw-Hill Encyclopedia of Science and Technology*, 1987, Vol. 1, pp. 602–605.

Peterson, Ivars. Pumping Gas to Fuel a Galaxy's Active Core. *Science News*, September 2, 1989, p. 150.

Price, Richard H., and Kip S. Thorne. The Membrane Paradigm for Black Holes. *Scientific American*, April 1988, pp. 69–77.

Punsly, Brian, and Ferdinand V. Coroniti. Relativistic Winds from Pulsar and Black Hole Magnetospheres. *Astrophysical Journal*, February 20, 1990, pp. 518–35.

Reading God's Mind. *Newsweek*, June 13, 1988, pp. 56–59.

Rich, R. M., et al. Luminous M Giants in the Bulge of M31. *Astrophysical Journal Letters*, June 15, 1989, pp. 51–54.

Robinson, Leif J. The Black Heart of the Milky Way. *Sky & Telescope*, August 1982, pp. 133–35.

Rothman, Tony, and George Ellis. Has Cosmology Become Metaphysical? *Astronomy*, February 1987, pp. 6–22.

Sambursky, Shmuel, editor. *Physical Thought from the Presocratics to the Quantum Physicists*. New York: Pica Press, 1974.

Scherrer, Robert. Our Cosmic Horizons, Part 1: From the Cradle of Creation. *Astronomy*, February 1988, pp. 4–45.

Seeing to the Edge of the Universe. *Discover*, April 1988, pp. 14, 16.

Segre, Emilio. *From Falling Bodies to Radio Waves*. New York: W. H. Freeman and Company, 1984.

Shapiro, Stuart L., and Saul A. Teukolsky. *Black Holes, White Dwarfs and Neutron Stars: The Physics of Compact Objects*. New York: John Wiley & Sons, 1983.

Skinner, G. K., et al. Localization of One of the Galactic Centre X-Ray Burst Sources. *Monthly Notes of the Royal Astronomical Society*, March 1, 1990, pp. 72–77.

A Smaller Milky Way. *Sky & Telescope*, August 1986, p. 129.

Smith, David H. Black Hole Weighs in at One Billion Suns. *Sky & Telescope*, August 1984, pp. 127–28.

———. A Sideways Look at Galactic History. *Sky & Telescope*, August 1986, pp. 111–114.

Smith, Robert W. The Great Debate Revisited. *Sky & Telescope*, January 1983, pp. 28–29.

Smothered Pulsars or Black Holes? *Sky & Telescope*, October 1988, pp. 340–41.

Sofue, Yoshiaki. Jet from the Milky Way Center: The Galactic Center Lobe. *Canadian Journal of Physics*, April 1986, pp. 527–30.

Sofue, Yoshiaki, and Toshihiro Handa. A Radio Lobe over the Galactic Center. *Nature*, August 16, 1984, pp. 568–69.

Sponge Universe? *Sky & Telescope*, December 1986, p. 574.

Stahler, Steven, and Neil Comins. The Difficult Births of Sunlike Stars. *Astronomy*, September 1988, pp. 22–33.

Starburst Galaxies: More Heat than Light. *New Scientist*, May 14, 1987, pp. 46–49.

Sternheim, Morton M., and Joseph W. Kane. *General Physics*. New York: John Wiley & Sons, 1986.

Taubes, Gary. The Great Annihilator. *Discover*, June 1990, pp. 69–72.

Thomsen, Dietrick E. Falling into a Black Hole. *Science News*, July 19, 1986, p. 40.

———. Farthest Galaxy Is Cosmic Question. *Science News*, April 23, 1988, pp. 262–63.

———. Watching a Black Hole's Diet. *Science News*, December 6, 1986, p. 359.

Thorne, Kip A., Richard H. Price and Douglas A. Macdonald, editors. *Black Holes: The Membrane Paradigm*. New Haven, CT: Yale University Press, 1986.

Townes, Charles M., and Reinhard Genzel. What Is Happening at the Center of Our Galaxy? *Scientific American*, April 1990, pp. 46–55.

Trefil, James. Galaxies. *Smithsonian*, January 1989, pp. 36–51.

Trimble, Virginia. Our Cosmic Horizons, Part 2: The Search for Dark Matter. *Astronomy*, March 1988, pp. 18–23.

Uchida, Keven, Mark Morris, and E. Serabyn. Study of AFGL 5376: An Unusual Extended Infrared Source near the Galactic Center. *Astrophysical Journal*, March 10, 1990, pp. 443–53.

Unravelling Fates of Black Holes. *Nature*, May 8, 1986, p. 111.

Vallee, J. P. Can the Large-Scale Magnetic Field Lines Cross the Spiral Arms in Our Milky Way Galaxy? *Astronomical Journal*, March 1988, pp. 750–54.

———. Magnetism of Spiral Galaxies. *Journal of the Royal Astronomical Society of Canada*, April 1984, pp. 57–74.

———. A Possible Excess Rotation Measure and Large-Scale Magnetic Field in the Virgo Supercluster of Galaxies. *Astronomical Journal*, February 1990, p. 459.

Valtonen, Mauri J., and Seppo Mikkola. The Problem of Few Black Holes. *Celestial Mechanics and Dynamical Astronomy*, 1989, vol. 46, pp. 277–85.

Verschuur, Gerrit L. Is the Milky Way an Interacting Galaxy? *Astronomy*, January 1988, pp. 26–31.

Waldrop, M. Mitchell. The Core of the Milky Way. *Science*, October 11, 1985, pp. 158–60.

———. Feeding the Monster in the Middle. *Science*, January 27, 1989, p. 478.

———. Heart of Darkness. *Science*, February 19, 1988, p. 866.

———. A Hole in the Milky Way. *Science*, May 21, 1982, pp. 838–39.

———. New Mysteries at the Galactic Center. *Science*, November 8, 1985, pp. 652–53.

Weaver, Jefferson Hane. *The World of Physics: A Small Library of the Literature of Physics from Antiquity to the Present*, Volume 1, New York: Simon and Schuster, 1987.

Wyse, Rosemary F. G., and Gerard Gilmore. The Galactic Spheroid: What Is Population II? *Astronomical Journal*, May 1988, pp. 1404–14.

The Year 10^{111}. *Scientific American*, January 1982, p. 80.

Yusef-Zadeh, Farhad, and Mark Morris. Summary of IAU Symposium No. 136 on the Galactic Center. *Comments on Astrophysics*, 1989, vol. 13, pp. 273–93.

Yusef-Zadeh, Farhad, Mark Morris, and D. Chance. Large, Highly Organized Radio Structures near the Galactic Center. *Nature*, August 16, 1984, pp. 557–61.

Zeilik, Michael. *Astronomy: The Evolving Universe*. New York: Harper & Row, 1985.

Index

Page numbers in italics indicate pages containing illustrations.

Milky Way Galaxy. *See also* Galaxies
  center of, 248–82, *275, 277, 279, 280*
  CO map of, 231
  corona of, 198–99
  discovery of shape, 83, 86–89
  "missing mass" of, 197–207
  origin of name, 23
  sun's location in, 231
  spiral arms of, 218, 239–49
  structure of, 191–94
Minkowski, Hermann, 112
MK system (luminosity classification),
  164–65
Molecular hydrogen, 227–28
Month, lunar, 45–46
Moon, draconic and sidereal periods of,
  46
Morley, Edward, 91, 101–3
Morris, Mark, 202, 262, 273–74, 276
Moskowitz, Bruce, *206*
Motion, laws of (Newton's), 70–73, 78,
  118
Mount Wilson Observatory, 129, 253
"Mule, The" (Asimov), 7
Music, mathematical principles of,
  40–*41*
Mythology, role in cosmologies, 19–35

Naked singularities, 141, 150
NASA
  Gamma Ray Observatory, 260
  Jet Propulsion Laboratory, 259
Navigation, celestial, by ancient Greeks,
  37
Nebulas
  bright or emission, 219
  reflection, 221
  spiral, 128
Neihardt, John, 25
Neptune, discovery of, 80
New Astronomy, The (Kepler), 66
Newman, E. T., 153
Newton, Sir Isaac, 74–82
  laws of motion, 70–73
  laws of universal gravitation, 65–66
Neugebauer, Gerry, 253–54
Neutrinos, 202–3
Neutron Star (Niven), 9

Neutron stars, 184–85
Niven, Larry, 9–13
Novae, 182–83
"Now You See It—" (Asimov), 7
Numbers and numerology, 32–33, 40

Odyssey (Homer), 37
On the Heavens (Aristotle), 50
Oort, Jan, 197–98, 235
Oort limit, 197
Open clusters (of stars), 170–71
Orion arm (of Milky Way), 239
Orion Molecular Cloud (OMC-1), 222
Orion Nebula, 89, 220–24
Osiander, Andrew, 62
Ostriker, Jeremiah, 198

Parallax, interstellar, 50–*51*, 188
Parsec, 188
Paschan series, 98
Pauli, Wolfgang, 202
Peebles, P. J. E., 198
Pendulums, Galileo and, 73
Penrose diagrams, 146, *147*
Penrose, Roger, 141
Pericles, 43
Perigalacticon, 209
Perseus arm (of Milky Way), 239, 243
Phaedrus, The (Plato), 44
Philolaus of Crotona, 41
*Philosophiae Naturalis Principia
  Mathematica* (Newton), 77
Philosophical Transactions (Royal
  Society of London), 131
Photoelectric effect, 105–6
Photino, 205
Photons, 81, 96, 106
Photon sphere (of black holes), 145,
  153
Physics (Aristotle), 49
Physical Review Letters, 150
Pima tribe, 23–25
Planck, Max, 2, 104–6
Planck's constant, 105
Planets
  identification of, 32–33
  motion of. *See also* Celestial motion
  Eudoxus's theory of, *46–48*